Local Distribution Pipelines in Nontechnical Language

LOCAL DISTRIBUTION PIPELINES
IN NONTECHNICAL LANGUAGE

Thomas O. Miesner

Disclaimer: The recommendations, advice, descriptions, and the methods in this book are presented solely for educational purposes. The author and publisher assume no liability whatsoever for any loss or damage that results from the use of any of the material in this book. Use of the material in this book is solely at the risk of the user.

Copyright © 2017 by
PennWell Corporation
1421 South Sheridan Road
Tulsa, Oklahoma 74112-6600 USA

800.752.9764
+1.918.831.9421
sales@pennwell.com
www.pennwellbooks.com
www.pennwell.com

Marketing Manager: Sarah De Vos
National Account Executive: Barbara McGee Coons

Director: Matthew Dresher
Managing Editor: Stephen Hill
Production Manager: Sheila Brock
Production Editor: Tony Quinn
Book Designer: Susan Ormston
Cover Designer: Elizabeth Wollmershauser

Library of Congress Cataloging-in-Publication Data

Names: Miesner, Thomas O., author.
Title: Local distribution pipelines in nontechnical language / Thomas O. Miesner.
Description: Tulsa, Oklahoma, USA : PennWell Corporation, [2017] | Includes bibliographical references and index.
Identifiers: LCCN 2016025555 | ISBN 9781593703776
Subjects: LCSH: Gas distribution--Popular works. | Natural gas pipelines--Popular works.
Classification: LCC TP757 .M54 2016 | DDC 665.7/44--dc23

All rights reserved. No part of this book may be reproduced,
stored in a retrieval system, or transcribed in any form or by any means,
electronic or mechanical, including photocopying and recording, without the
prior written permission of the publisher.

Printed in the United States of America

1 2 3 4 5   21 20 19 18 17

contents

Acknowledgments ix

1 How Pipelines Differ. 1
    Pipelines Are Ubiquitous 1
    Energy Pipelines 1
    Pipeline Categories 2
    From Producer to Consumer. 3
    Summary. 14

2 Two Beginnings 15
    Light and Heat: Safety and Security. 16
    Gas from Coal. 16
    Enter Natural Gas 18
    More about Manufactured Gas 19
    Early Beginnings: 1800 to 1850 21
    Powering the Industrial Revolution and Moving into the Home:
       1850 to 1900 23
    Glory and Fall of Manufactured Gas:
       1900 to 1950 30
    Turbulent Times to Shining Star: 1950 to 2000 42
    Summary. 46
    Notes. 47

3 How Pipelines Work 51
    The Physics of Fluid Flow 51
    Municipal Water Systems 53
    Friction Losses, Pipe Lengths, and Flow Rates 54
    Hydraulic Properties of Hydrocarbon Fluids. 59
    Hydraulics. 67
    Summary. 76
    Note. 77

4 Components and Equipment 79
Introduction 79
Pipe 80
Coatings 86
Fittings 91
Valves 98
Actuators 105
Meters 109
Meter Provers 116
Odorant Skids 117
Rectifiers 117
Anodes 118
Storage 119
Other Components 121
Summary 121
Notes 123

5 Natural Gas Transmission Line Operations 125
Natural Gas Lines: A Brief Review 128
Central Control Rooms 146
Summary 157

6 Local Distribution Pipeline Operations 159
Local Distribution Pipelines: An Overview 160
Functions and Tasks 172
Operations 173
Customer Service 180
Gas Supply 184
Other Groups 189
Summary 190
Notes 191

7 Liquefied Natural Gas 193
What Is LNG? 194
A Brief History of LNG 194
Why Liquefy Natural Gas? 195
The Science behind LNG 196
Gas → Liquid → Gas 196
Export and Import Terminals 202
Other LNG Facilities 203

	Summary..203
	Notes..204
8	**Releases, Leaks, and Leak Management**.......................205
	Introduction...205
	Releases...207
	Leaks..208
	Detecting Leaks..21z
	Investigating Leaks......................................218
	Locating Leaks...218
	Leak Management..220
	Summary..221
	Notes..222
9	**Asset Integrity**..223
	Introduction...223
	Risk...224
	Mining Past Data...227
	Failure Mechanisms and Forces............................231
	Preventing Releases......................................236
	Integrity Management Plans...............................242
	Determining Asset Condition..............................244
	Defects and Failures.....................................245
	Repairing Defects and Failures...........................246
	Summary..255
	Notes..249
10	**Control Systems and SCADA**................................251
	Control Systems..253
	SCADA..255
	Design and Control.......................................257
	Summary..274
11	**Design and Engineering**...................................277
	Natural Gas Engineering Functions........................278
	Standards and Codes......................................279
	The Engineering and Design Process.......................280
	Design and Engineering...................................286
	Pipeline Design..286
	Facility Design..296

Facility Modifications...308
Power System Design..308
Control System Design 309
Other Engineering Tasks 309
Summary..309
Notes..310

12 Construction 311
Introduction...311
Contracting..312
Permitting...313
Plastic Lines and Connections 314
Trenchless Construction Techniques 320
Steel Lines..323
Facility Construction 325
Quality Control 327
Safety 331
Data Collection 331
As-Builts and Inventories 331
Operating Procedure Manuals 332
Summary..332

13 Business Models and Expenditure Decisions 335
Ownership..335
Bundled, Unbundled, and Open Access..........................336
Revenues...337
Expenditure Decisions..340
Valuation..344
Summary..346
Notes..347

14 Challenges for the Future 349
Public Safety..350
Reliability..354
Efficiency...358
Environmental Performance 359
Aging Workforce 359
Summary..360
Notes..360

Index 363

acknowledgments

In 2006 PennWell released *Oil and Gas Pipelines in Nontechnical Language*, which I coauthored with Bill Leffler. Without Bill's encouragement and tutelage, the transmission book, and by extension, this book, would never have been written. My thanks to Bill for his patience, encouragement, wit, and most of all his teaching.

The transmission book stopped at the city gate, and it occurred to me the natural gas pipeline story lacked the final link—local delivery.

My deep background is the hazardous liquids pipeline business—crude oil, refined products, and the like. While liquids and gas obey the same laws of physics, parts of the businesses and operations have interesting differences. Fortunately, working on oil and gas pipeline industry common issues while at Conoco brought me into contact with several natural gas transmission industry professionals. When I told them of my plans to write the first book, they enthusiastically gave me free reign to learn from their people. Making the rounds with natural gas transmission technicians and control room operators, I learned the natural gas transmission industry, but I realized there was still much I had to learn before I could write knowledgeably about the distribution side of the business.

Shortly after finishing the transmission book, I was blessed to meet Steve Vitale, who spent many years in an assortment of technical and engineering roles in the natural gas distribution business. Steve patiently explained the business to me and answered questions over the ensuing years as the book progressed. He even wrote the introductory story for the LNG chapter.

Three Davids provided valuable assistance with this book. David Waples, author of *The Natural Gas Industry in Appalachia: A History from the First Discovery to the Tapping of the Marcellus Shale* provided extensive source materials for chapter 2 and facilitated gaining permission to use various photos providing a visual history of the industry. David Klimas, senior vice president and chief engineer with EN Engineering, reviewed chapter 11, providing important upgrades to what I had written. Finally, David Vanderpool, principal, Vanderpool Pipeline Engineers Inc., whom I worked with at Conoco Pipeline for many years and with whom I wrote a chapter on pipeline engineering for another publisher, reviewed drafts of chapters 3 and 11, offering many excellent clarifications and improvements. David also generously allowed me to use some of his firm's engineering standards in chapter 11.

Through a series of litigation consulting cases, I met Mark Kazimirsky, now president of MK Consulting Services Inc. He has a deep background in controls and SCADA and provided valuable insight into those topics as relates to the LDC industry. Mark reviewed the draft of chapter 10, providing edits as he went.

Over the years, Cheryl Trench, principal, Allegro Energy Consulting has performed extensive review and analysis of natural gas distribution, natural gas transmission, and liquid pipeline incident data. When I had questions about PHMSA data or about one of the many reports she published, Cheryl always answered those questions and carefully put the data into context for me.

Leak detection and monitoring are critical issues, and I appreciate the assistance provided to me with them by Heath Consultants in the persons of Paul Wehnert and Vivian Marinelli. They spent time reviewing history, current practices, and leak survey equipment. They also paved the way for me to use various pictures from Heath in chapter 8.

Sometimes industry insiders are happy to comment but don't want acknowledgment, which is exactly what happened for chapter 8 and 14. She helped me understand important differences between transmission and distribution when it comes to leaks and release management as well as helping me think about challenges for the future as I wrote chapter 14.

A wide assortment of operating companies and professionals taught me about the local distribution business, patiently answering my questions, escorting me to their facilities, reviewing chapter drafts, and providing comments and suggestions. Unfortunately, in nearly all cases, their companies and organizations preferred that I not acknowledge them by name, which I understand and respect. Nevertheless, I am deeply grateful to these men and women who freely shared their knowledge and time. Without them this book would not exist.

Many people contributed to this book by sharing their knowledge, either teaching me or allowing me to learn by observation. Others took a keen interest and freely lent their expertise and entry to the inner workings of the industry and their company. It is impossible to thank each one by name. To all these people I express my thanks. Nonetheless, and as always, I had the last word, so I carry the burden of interpreting correctly what others offered.

Finally, I thank my family, and particularly my wife, for the patience, understanding, encouragement, and love she provided during this effort. As any well-married author will attest, those four gifts are the prerequisites of timely book completion.

chapter 1

How Pipelines Differ

Always do right. This will gratify some people and astonish the rest.

—Mark Twain

Pipelines Are Ubiquitous

In any city, in any developed country, the day begins. Families awaken and prepare for the day. Water arrives at the sink and shower via water pipelines. Sewage leaves via sewage pipelines. Natural gas arrives at the stove via an interconnected natural gas network. The school bus arrives, fueled by diesel that began as crude oil and was transported via a network of crude oil and refined products pipelines. Parents travel to work via public transportation or private cars, which also depend on fuels transported by pipelines. Once at work, employees use computers powered by electricity generated from natural gas. The natural gas is, in turn, supplied to power plants from the natural gas pipeline grid. No one really wants pipelines, but everyone needs them.

Energy Pipelines

Most people think pipelines are pipelines. All pipelines, whether gas, oil, water, or sewer, obey the same laws of physics but perform unique functions and operate differently. Put a group of oil pipeliners together with natural gas pipeliners, or even natural gas transmission pipeliners with natural gas local distribution pipeliners, and they might be speaking the same language, but with different dialects. Oil and gas pipelines (sometimes broadly referred to as *energy pipelines*) may look essentially the same, obey the same laws of physics, be installed in largely the same manner, and face the same regulatory and social challenges. However, that does not mean they use the same nomenclature or operate in the same manner.

There are critical differences between gas pipelines and oil pipelines. For instance, gas is compressible, meaning the volume occupied by a given number of molecules gets smaller when the pressure goes up and larger as the pressure goes down. In contrast, oil is a liquid, meaning the volume it occupies changes only slightly in response to changes in pressure. The relative compressibility of gas means gas pipelines use compressors, whereas oil pipelines use pumps, among many other operational differences.

Pipeline Categories

Energy pipelines are broadly classified by the commodity they transport, the function they provide, and whether they are located onshore or offshore. Commodities transported include the following:

- *Natural gas.* A mixture composed of primarily methane.
- *Oil.* Unrefined fluids such as crude oil and condensate, and refined products such as gasoline, diesel, jet, home-heating, and other finished fuels.
- *Natural gas liquids (NGLs).* Ethane, propane, normal butane and isobutene, pentane, and other heavier (larger) hydrocarbons, closely related to liquefied petroleum gas (LPG).
- *Chemical feedstocks.* Ethylene and propylene, often derivatives of NGLs.
- *Biofuels.* An emerging type; most notably at this time, ethanol.

Functions performed include the following:

- *Gathering.* Aggregating quantities for shipment.
- *Transmission.* Transporting relatively large amounts longer distances (also sometimes called *mainlines* or *trunk lines*).
- *Distribution.* Disaggregating for delivery to final destinations (including mains and service lines).

This book deals with natural gas local distribution lines. As an added bonus, a chapter is included that covers the basics of natural gas transmission lines, which are a crucial link in getting gas from production fields to the marketplace. However, before focusing exclusively on natural gas local distribution pipelines, a few words about the other forms of hydrocarbon pipelines are in order.

First cut liquid (oil) pipelines versus gas pipelines

Liquid pipelines include crude oil pipelines and refined products pipelines. Operations of some types of pipelines are more or less indistinguishable from oil pipeline operations, including chemical liquids pipelines, natural gas liquids pipelines, LPG pipelines, and probably biofuels. On the gas side, pipelines that move anhydrous ammonia, carbon dioxide, and gaseous chemicals like ethylene operate much like natural gas pipelines.

From Producer to Consumer

Both the oil and natural gas pipeline value chains begin in the *oil patch* or production field. Oil and gas are extracted from *reservoirs*, or locations where the hydrocarbons are found. A reservoir may occur where oil and gas collect after migrating from the source rock, or the source rock itself can be a reservoir if the source rock formation was tight enough and structurally capable of trapping the oil and gas.

Reservoir and source rock

Geologists and geophysicists analyze seismic data in their search for source rocks in which hydrocarbons have formed. After the oil and gas form, pressure pushes them toward the surface. Sometimes in the path of flow toward the surface, the hydrocarbons become trapped by impervious layers of rock, and a reservoir forms. For many years, production was only from reservoirs, because the technology did not exist to free oil and gas trapped in some source rock formations.

Due to technological advances, well engineers can now fracture (frack) or break apart the source rock. In this process, mixtures of water, sand, and various chemicals needed to prevent corrosion and fouling are pumped into the source rock formations under high pressure. The high-pressure water breaks the formation apart, and the sand lodges in cracks and crevices to keep the path open for the previously trapped oil and gas to flow. Fracking has opened (no pun intended) vast new resources of oil and gas.

Midstream

Pipelines are part of the *midstream* industry, forming the critical link between *upstream* (exploration and production) and *downstream* (processing and marketing) in the oil and gas value chain. Some of the other parts of the midstream industry include the following:

- Gas dehydration
- Gas treating
- Methane extraction plants
- Gas fractionation
- Gas storage
- Crude oil trucking
- Crude oil storage
- Refined products storage and loading terminals

Value chain is a concept popularized by, among others, Michael Porter. In his 1985 bestseller, *Competitive Advantage: Creating and Sustaining Superior Performance,* Porter discusses the chain of activities that add value to a product as it moves along its journey to market.

Pipeline value chain

Figure 1-1 shows the pipeline value chain.

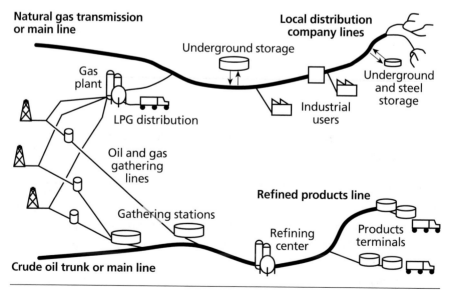

Fig. 1–1. Pipeline value chain

The pipeline value chain starts at the wellhead, where flow lines (generally called gathering lines in Canada) take the production stream from the well to production processing facilities. At the production processing facility, oil and gas are separated from each other and from everything else, and the oil and gas pipeline value chain splits (fig. 1-2).

Fig. 1–2. Production processing facility. Processing equipment, shown on the right, separates oil and gas from each other and from water and other contaminants. Oil is stored in the tanks, and the gas leaves the facility via pipeline.

Crude oil pipelines

At the production tank batteries, crude oil from one or more wells is aggregated. From there, gathering pipelines (generally called feeder lines in Canada) or specially designed trucks move it on to gathering stations (fig. 1-3).

As with most economic matters, the decision to connect crude oil tank batteries to pipelines is a balance of investment, cost, revenue, and increasingly, the ability to secure the permits to construct the pipeline.

Prolific wells and those located near existing gathering lines are normally connected directly to gathering lines and not trucked. Those producing less, or costing more to connect, are likely never directly connected. Whether transported by truck or pipeline, the crude oil is carefully measured and sampled as the pipeline company accepts custody from the producer. Figure 1-4 shows a *custody transfer unit*, where crude oil is measured and sampled to determine quality and sediment and water (S&W) content as it is being delivered by truck to the crude oil gathering station.

Fig. 1–3. Crude oil truck

Fig. 1–4. Crude oil automatic custody transfer (ACT) unit

Crude oil gathering lines normally consist of pipe from 2 inches to 12 inches in diameter. They originate at production tank batteries and end at locations along crude oil main lines, trunk lines, or transmission pipelines. Normally 8 inches or more in diameter, some *crude oil main lines* begin near crude oil production fields, receiving more and more crude oil as subsequent gathering lines inject their volumes. The crude oil can be injected into the continuous flow of crude oil as it goes past. If the crude oil is sufficiently different from that flow, the pipeline can be stopped upstream of an *injection point*, and a volume of crude can be pumped in as a separate *batch*. The batch is then tracked as it moves down the pipeline and is delivered to separate customers or into segregated tanks at the destination. Rather than receiving directly from gathering lines, crude oil main lines sometimes originate at *crude oil market centers*, where they receive crude oil from other main lines.

Regardless of where the crude oil originates, the role of a crude oil mainline is to deliver crude oil to refineries (fig. 1–5).

Fig. 1–5. Refinery

Some crude oil lines deliver directly to refineries, while others may deliver to market centers like Cushing, Oklahoma, for storage and subsequent movement to refineries. Still other crude oil lines move crude oil to seaports like Ceyhan, Turkey, where the crude oil is loaded onto tankers for delivery to other countries (fig. 1–6).

Fig. 1–6. Loading a tanker at a marine export facility

Refined products pipelines

The refined products pipeline value chain begins at refineries and ends at *petroleum products terminals,* collections of large tanks along the pipeline, located near consumers (fig. 1–7).

Fig. 1–7. Refined products terminal. The storage tanks are shown on the left, and the truck loading rack is shown on the far right.

Products move down the pipeline in batches. Sometimes the entire flow of the pipeline is diverted into a terminal tank; at other times, only part of the stream is diverted into the terminal, while the rest travels past, destined for locations farther downstream.

At the terminal, gasolines and diesel fuels are loaded onto trucks and sometimes railcars for delivery to retail outlets or commercial and industrial consumers (fig. 1-8).

Fig. 1–8. Refined product truck loading facility. Load arms are connected to a manifold on the bottom of the tank to facilitate loading different products into various compartments on the truck.

The network at either end of a product pipeline can vary. For example, the Yellowstone Pipeline originates in Billings, Montana, where it receives batches of product from only three refineries. The pipeline moves the product to terminals across Montana. Yellowstone is the single source for those terminals. In contrast, the Explorer Pipeline begins on the US Gulf Coast, where it receives batches of product from a dozen or more refineries and other pipelines. When it reaches Dallas, Tulsa, St. Louis, and Chicago, Explorer competes with other pipelines coming into those areas. It also competes with other refineries in the area.

Natural gas pipelines

Natural gas, like crude oil, is aggregated by gathering lines. Smaller lines deliver into larger lines, normally through meters (fig. 1-9).

Fig. 1–9. Natural gas delivery from one gathering line into another through an orifice meter.

Unlike crude oil, natural gas does not go into tankage following separation. It is kept under pressure as it travels through successively larger diameter gathering lines. Sometimes, depending on gas quality and contaminants, the natural gas is injected directly into gas transmission lines without processing.

Raw gas, the natural gas stream prior to gas plants, contains more than just methane (natural gas). In addition, it contains water and other heavier (larger) hydrocarbon molecules, or the so-called natural gas liquids or NGLs. It also contains contaminant gases. In successive size order, NGLs are comprised of ethane, propane, butane, and other heavier molecules. Contaminant molecules can include carbon dioxide, nitrogen, oxygen, hydrogen sulfide, and other elements like argon, helium, neon, and xenon. If left in the gas stream, these contaminant gases can combine with water to form acids, most notably carbonic and sulfuric, thereby speeding internal corrosion. Accordingly, the gas stream may undergo dehydration to remove water. It may also be processed in an amine plant to remove carbon dioxide and hydrogen sulfide. The specifics of gas treating depend on the gas stream characteristics, which in turn depend on the geology from which the gas was produced.

Prior to delivery into the transmission line, and at the end of the gathering system, the raw gas stream is delivered to a gas processing plant. At the plant, ethane, propane, butane, and other heavier molecules are removed from the gas stream, leaving nearly all methane molecules. The heavier molecules that are removed from the methane may go through a fractionation process, where they are split into their individual components (fig. 1–10).

Fig. 1–10. Natural gas processing plant. This plant also fractionates the NGLs removed from the methane.

Upon removal, the natural gas liquids are sold primarily into the chemical and LPG fuel markets. They move from the gas plant via tank cars, trucks, or their own pipeline.

Gas plants typically are located next to, or at least near, gas transmission lines, into which they deliver the methane stream. Compressors, either positive displacement or centrifugal, add pressure as the gas enters the transmission line to push the gas stream along (fig. 1–11).

Fig. 1–11. Natural gas positive displacement compressor

Compressor *booster stations* are installed along the line about every 50 or so miles to add back the pressure lost due to the gas molecules rubbing against the pipe wall and each other as they move along. Since the processed gas is of a relatively consistent quality, batching is not required on gas transmission lines. Lack of batching means one natural gas transmission line can deliver into another at any time if it has sufficient pressure and shippers have made the necessary arrangements. These deliveries are made through interconnects or hubs (fig. 1–12).

Fig. 1–12. Natural gas interconnect. Interconnects between multiple pipelines at one location are often called *hubs*.

Natural gas transmission lines also deliver to local distribution mains at city gates, also called *town gates*. Formerly, natural gas transmission lines delivered only to local distribution companies (LDCs) for further movement to homes and businesses. However, due to regulatory changes, natural gas transmission lines increasingly deliver directly to large end users like industrial plants, businesses, commercial sites, and power generation plants, bypassing LDCs. Sometimes, instead of going directly to consumers, natural gas is stored for future use. The storage can be in depleted oil or gas fields, underground caverns or aquifers, or less frequently aboveground steel vessels in either compressed of liquefied form. Natural gas local distribution lines receive natural gas from transmission lines and deliver it to homes, businesses, schools, hospitals, factories, and the like (fig. 1–13).

Fig. 1–13. Natural gas delivery meter to a home

Summary

- Pipelines in developed countries are ubiquitous.
- Recent technological improvements allow production from source rock in addition to reservoirs.
- One classification for gas and oil pipelines is based on the commodity they transport, whether oil or natural gas.
- Another classification is based on the function of the pipeline, such as gathering, transmission, or distribution. Distribution is further divided into mains and service lines.
- The oil pipeline value chain starts at the production tank battery or field separation facility and extends to the refined products terminal.
- The natural gas value chain starts at the production battery or field separation facility and extends to the ultimate consumer.
- Natural gas transmission pipelines interconnect at hubs or market centers.
- Natural gas transmission pipelines deliver natural gas to local distribution pipelines, which in turn deliver it to the end users.

chapter 2

Two Beginnings

Begin at the beginning . . . and go on till you come to the end: then stop.

—Lewis Carroll

The first discovery of most things gets lost in lore and legend, and so it is with the history of natural gas. In one account, a Greek goat herder around 1000 BC observed that the goats became giddy when they congregated around a certain spot. The herder reported this phenomenon to the authorities, who proceeded to investigate. The authorities also became light-headed at the location, which they concluded was due to spirits living in the area. Accordingly, they had a temple constructed on the site and installed a priestess known as the Oracle of Delphi. Religious leaders in India, Persia, and Greece continued to capitalize on the supposedly supernatural aspect of burning gas seeps that were accidentally ignited by lightning. Recent research reported in *National Geographic* now supports this story.[1]

Not until about 500 years later did the Chinese turn gas seeps to productive use. Legend has it they hollowed out bamboo shoots, connected them together with clay, and "piped" the gas short distances. This allowed them to evaporate water from brine and collect the remaining salt.

Both George Washington and Thomas Jefferson noted in their personal journals encounters with naturally occurring gas seeps similar to those found by the Greeks and the Chinese. In 1775 George Washington even tried to get one square mile in the Kanawha Valley of West Virginia set aside so the public could view burning gas issuing from underground.[2] But until the late 18th century, the technological challenges of capturing, transporting, and safely burning gas—natural or otherwise—kept it from widespread use.

Light and Heat: Safety and Security

Until the mid-19th century, lanterns or candles were used for lighting. Fuel for lighting came from many sources, including animal fats (called *tallow*), waxes of various types, and oils, such as whale, olive, and fish. Heat came from coal, wood, and other combustibles, including rice straw and cow dung.

Town leaders, charged with the safety of their citizens, searched for ways to effectively provide street lighting as a deterrent to crime. In 1416 Sir Henry Barton, the Lord Mayor of London, ordered homeowners to hang lights out on the street sides of their houses from All-Hallows to Candlemas (then considered the beginning of spring).[3]

Enterprising souls in search of fame, fortune, or simply the joy of discovery investigated better means of producing heat for cooking and comfort, and light for security, work, and leisure. Their searches also led at the beginning of the 19th century to the use of gas, but from two different sources, manufactured and natural. *Manufactured gas* or *town gas* was produced at first from coal and wood and then later from pitch, tars, and even oil. It was extensively used up until about the time of World War II—and in some locations far longer.

Gas from Coal

With most historical technological advances, such as cars, airplanes, and the electric light bulb, to mention a few, many people were striving diligently to get the technology to the point of commercial potential. However, usually only one person received the credit. In the case of manufactured gas, the person largely credited with its discovery is William Murdoch (or Murdock) (1754-1839). In his youth, Murdoch is said to have put chunks of coal in his mother's teapot, welded it shut, and lit the gases that escaped from the spout when the pot was heated. The energetic Scottish "engineer" was employed by James Watt (the inventor of the steam engine) and Mathew Boulton in their Soho Foundry Steam Engine Works in Birmingham, England. Murdoch became James Watt's right-hand man, keeping the steam engines running.

Even though constantly busy with steam engines, Murdoch still found time to continue his early "coal gasification" experiments. Around 1792 Murdoch placed coal into an iron retort in his backyard, heated it, and piped the resultant gases about 70 feet into his living room. Placing a match at the end of the tube, he was rewarded with a pleasing (although smelly) flame, vastly superior in lighting ability to anything previously produced.[4]

Competition, regulation, and self-promotion—from the beginning

In 1796 Murdoch resumed his work at the Soho Foundry at Birmingham, focusing his efforts on development of commercial gas lighting. By 1798 Murdoch had illuminated the main building of the Soho Foundry with gas, and in 1802 he lighted the outside of the building with open gas flares, amazing the public.

In 1801 Murdoch gained an assistant, Samuel Clegg (1781–1861), who quickly grasped the importance of this new discovery. Clegg subsequently left Boulton & Watt to work for the competition. In 1805 both Murdoch and Clegg built gas plants to illuminate cotton mills. The former assistant, Clegg, finished his work at Henry Lodge's Hebden Bridge mill two weeks ahead of Murdoch's lighting at Phillips & Lee's mill in Salford. Clegg's gas plant thus gained recognition as the first commercial coal-gas plant.[5] Murdoch continued to build gas plants to support individual factories.

Clegg partnered with Fredrick Albert Winsor (1763–1830), a German who apparently learned about gas lighting from Philippe LeBon (1767–1804). LeBon, a Frenchman, experimented with and received a patent for producing combustible gas from sawdust. He was never able to solve the terrible odor problems associated with this process. Winsor, who was better at public relations than science, moved to London in 1803 and began lecturing on the topic of gas lighting. He captured the imagination of venture capitalists of the day and used their money to secure premises on Pall Mall in London. There he had demonstration equipment installed, including two carburizing gas furnaces.

LeBon had the furnaces connected to lights installed along Pall Mall, using sheet lead bent into a cylindrical form and soldered at the edges. This initial demonstration of gas lighting amazed the residents, and some of them remained along the mall until nearly midnight. After more demonstrations, in November 1814, Winsor's company received permission to install mains along Pall Mall and other streets to supply lighting.[6] By 1819 London had 288 miles of pipe supplying 51,000 burners. By 1830 most of the larger cities and some towns in Britain had gas street lighting in one form or another.

Meanwhile in the United States

One of the first recorded uses of manufactured gas was by the Italian fireworks makers, M. Ambroise & Company of Philadelphia, Pennsylvania. In 1796 the company lit elaborate and fanciful chandeliers with coal gas at a public gathering in a local amphitheater. In 1802, they used illuminating gas in a sideshow at the Haymarket Gardens in Richmond, Virginia.[7] Later David Melville (1773–1856) received the first US gaslight patent on March 24, 1810.[8] Only a few years after that, in 1817, the Baltimore City Council granted the Gas Light Company of Baltimore a contract "to open a street or streets for the purpose of laying pipes for trying the experiment and estimating the expense at which it could contract with the Council for lighting the city."[9] Gas plants sprang up, and by 1830 many cities in the United States had adopted gas street lighting.

Enter Natural Gas

In 1821, a Fredonia, New York, blacksmith, William Aaron Hart, pounded into the ground a few lengths of pipe in the vicinity of some strangely ignitable gas bubbling from a running creek bed. He fashioned a floating, open-bottomed box next to his well and captured the gas as it surfaced. Finally, he ran hollowed-out logs as pipes to a Lake Erie lighthouse a half mile away, and supplied several businesses in town through small diameter lead pipes. The lighthouse beacon and chandeliers of the theater in town lit up, and Hart had created the natural gas pipeline industry. Early natural gas drilling was primitive, often conducted with a spring pole drill (fig. 2–1).

Fig. 2–1. Spring pole drilling
Courtesy: National Fuel Gas.

More about Manufactured Gas

Quite a number of gas manufacturing methods were eventually discovered and employed, but two of them, coal carbonization and carbureted water gas, comprised 85% to 90% of total production.

Coal carbonization

Employed by William Murdoch, this process heated coal in a closed vessel called a *retort*. The closed vessel kept oxygen away from the coal. Consequently, the coal did not burn but rather released its volatile compounds, leaving *coke*, which was composed of carbon and impurities. (Afterwards, the coke was often burned to heat the retort.) Passing from the retort, the gas stream went through several other vessels prior to going to storage or to the marketplace (fig. 2–2).

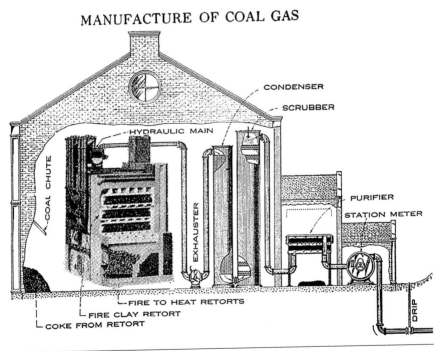

Fig. 2–2. Coal gas manufacturing. This illustration reportedly first appeared in Samuel S. Wyer's *The Smithsonian Institution's Study of Natural Resources Applied to Pennsylvania's Resources: Based on Latest Governmental Data* (1922).

Gas manufactured by this process consisted of a mixture of mostly methane, ethylene, carbon monoxide, and carbon dioxide. The ethylene gave it sufficient luminescence that the lamplights needed no mantle to burn brightly, an advantage over natural gas, which contains almost no ethylene.

Carbon monoxide is highly toxic, and fatalities with manufactured gas tended to be due to its carbon monoxide content. In contrast, the toxicity of natural gas is low. Fatalities related to natural gas usage tend to occur either due to oxygen deprivation or from a resulting explosion when a spark ignites the confined natural gas.

Carbureted water gas

In 1873 the carbureted water gas process was developed by T. S. C. Lowe. This process consisted of passing high-pressure steam over hot coal. Water reacted with carbon monoxide to produce hydrogen and carbon dioxide. The improved efficiency provided a great boost to the manufactured gas industry as the world entered the 20th century (fig. 2–3).

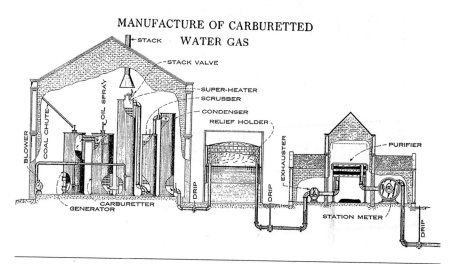

Fig. 2–3. Carbureted gas manufacturing. This illustration reportedly first appeared in Samuel S. Wyer's *The Smithsonian Institution's Study of Natural Resources Applied to Pennsylvania's Resources: Based on Latest Governmental Data* (1922).

Depending on the process and the coal or oil used, different qualities of fuel were produced. Each of these two processes, and several others, were modified and improved over time. They all produced by-products, including coke, tars, and pitches. Some of these by-products still haunt old plant sites to this day.

Early Beginnings: 1800 to 1850

Discovering how to manufacture gas, or find it and produce it, was just the beginning. Now it had to be delivered to market. Little was known about gas and its behavior in the early 1800s. As an example, pipes carrying gas were thought to be hot, and the architect designing manufactured lighting for the House of Commons insisted the gas lines be placed four to five inches from the walls so the gas going through the pipes would not catch the walls on fire.

Pipes were hard to come by, and musket barrels were sometimes used, with the muzzle of one being screwed into the breach of the next (fig. 2–4).[10]

Fig. 2–4. Manufacturing service pipe from gun barrels
Source: Drawing from Oscar E. Norman, *The Romance of the Gas Industry* (Chicago: A.C. McClurg & Company, 1922).

Some manufacturers at first refused to risk their time and money to produce the pipelines needed for such a "foolish and unlucky thing as gas." Over time multiple piping materials such as lead, tinned steel, asphalt, cement, concrete, and even wood were tried (fig. 2–5).

Fig. 2–5. Hollowed-out log used to transport gas. Note the iron bands added to reinforce the wood.

Cast iron pipe was first used in Langensalza, Germany, around 1562, according to the Cast Iron Soil Pipe Institute.[11] It eventually emerged as the material of choice for mains. Necessity continued to drive invention, and Aaron Manby of the Horseley Iron Works discovered how to weld longitudinal seams with machines rather than manually. This improved manufacturing method led to reduced costs and higher quality, especially for the smaller diameter service lines.

Lack of dependable meters meant gas volumes were sometimes estimated and sold by contract. Gas suppliers hired inspectors, who visited the houses of consumers and shut off the gas at the street cocks when they found gas lights still lit after the appointed contract hour. All sorts of disagreements arose under this lump-sum method as gas companies experienced large losses from customers abusing the system.[12] In 1815 the gas meter and the *governor*, a device to regulate pressure, were invented by Samuel Clegg, Sr.[13] Metering improvements followed as gas companies and customers sought fair means of buying and selling gas.

The new lighting proved irresistible to consumers, and by about 1826 most major cities and towns in the United Kingdom had gas manufacturing plants for street and factory lighting. In 1824 the Imperial Continental Gas Association was founded by Sir Moses Montefiore and some of his colleagues to establish gas works on the Continent. Hanover and Berlin received manufactured gas in 1825 and 1826, respectively. By 1850 there were a number of gas lighting companies in the United States, including ones in Baltimore, Boston, New York City, Louisville, Kentucky, St. Louis, Cincinnati, Philadelphia, Newark, Charleston, New Haven, Providence, and Chicago.[14] Manufactured gas was the predominant gas, with natural gas limited to a few towns located near natural gas production.

Just about this same time a new competitor entered the scene as kerosene was distilled by Dr. Abraham Gesner in 1846.[15] Kerosene competed against gas for home and street lighting for about 40 years, until both were replaced by electricity following Edison's invention of the light bulb in 1875.

Powering the Industrial Revolution and Moving into the Home: 1850 to 1900

In the second half of the 19th century, the use of gas expanded from lighting streets, factories, and a few homes into a source for home cooking and heating, as well as power for factories. Two key inventions moved the industry forward. The first was the Bunsen burner, created by Robert Bunsen in 1855, which significantly improved heating efficiency. The second was the gas mantle, developed by Carl Auer in 1887, which provided brighter illumination with less gas usage. By the end of the 19th century, there was at least one gas plant on each of the inhabited continents.

Multiple stove, oven, and water heater designs were developed and promoted during this same time period. The Tyson Engine Company of Philadelphia even manufactured about 100 gas-driven steam engines for powering sewing machines.[16] Gas also made significant inroads into all sorts of industries, including baking and candy making, coffee roasting, smoking meat, pasteurizing mills, glass drying, and china decorating. Gas was employed in hat shaping, shoe drying, clothes pressing, cigar lighting, barber boilers, vulcanizing, lumber drying, tinning, varnish boilers, rivet heaters, forging, brass melting, galvanizing, coloring and rust-proofing metals, and welding and cutting metals, among other industries.[17]

Sulfurous coal fumes belching from hundreds of factories caused Pittsburgh to be named the Smoky City around 1870. The pollution spurred the conversion to the use of natural gas from nearby gas reserves. Recognizing the clean-burning nature and lower cost of natural gas, entrepreneurs began promoting it for factory usage. The earliest recorded use of natural gas for steel making is around 1870 at the Great Western

Iron Company on the Allegheny River.[18] Factory use of natural gas grew throughout the remainder of the decade.

In 1883 George Westinghouse (founder of Westinghouse Electric) decided to have a gas well drilled in his backyard. As luck would have it, the drillers hit a large gas deposit. Westinghouse acquired the abandoned Philadelphia Company, which had been chartered in 1870 by the Pennsylvania legislature to produce and distribute natural gas. By 1887 the Philadelphia Company served about 5,000 residential and 470 industrial companies. The average charge to heat a typical shop was $2.50 per month.[19] Natural gas was positioned to make inroads but needed additional transmission technology before it could expand beyond local markets.

Natural gas (composed primarily of methane) has about twice the energy content per cubic foot as manufactured gas (composed of a variety of other gases). Mixing the two in the same distribution system could create safety problems if the higher energy gas entered appliances designed to burn the lower energy gas. Consequently, mixing the two gases in the same system happened infrequently, at least after the first several attempts.

Distribution

High pressure during this time was 100 psi, with most distribution lines operating at less than 15 psi. Cast iron and wood were used for mains, and wrought iron with threaded fittings was the material of choice for service connections. Some accounts discuss gas losses in the range of 10% to 30%, primarily from poor metering and less-than-tight connections. These losses were costly, and the distribution function, up to now a necessary part of the business but second fiddle to manufacturing, moved into an industry of its own. Distribution engineers developed handbooks, charts, and graphs to simplify calculations and began to standardize the layouts of mains and service lines as they sought to improve efficiency and reduce leaks.

Business and regulations

The lack of technology needed to move gas long distances meant gas manufacturing and distribution were local enterprises. A patchwork of private and municipally owned manufacturing and distribution companies emerged. Sometimes they competed with each other, and sometimes they were given exclusive charters and contracts for city lighting. By 1847, the United Kingdom saw the need to more closely regulate the enterprises and passed the Gas-Works Clauses Act. Among other things, this act gave gas companies the right to lay pipes under the streets and established the regulations they must follow to do so. This act was followed by the Sale of Gas Act in 1859, the Metropolis Gas Act of 1860, and the City of London Gas Act of 1886, to mention a few. Regulations in the United States and other countries followed suit.

In 1860, Samuel Adams Beck built a gas works in the United Kingdom that consumed 1 million tons of coal annually and produced 1.75 million cubic meters of gas daily. By 1880, 30,000 men were employed at the works, and it covered an area larger than the city of London.[20]

Elsewhere in Europe, plants sprang up to meet city needs, often funded by English companies. Holding companies with limited regulations were the order of the day. By 1900, Germany reportedly was producing 1.2 billion cubic meters of gas per year. By 1900, manufactured gas plants were located on every continent except Antarctica. However, the vast majority of gas usage, along with nearly all the regulations relating to its manufacture, transportation, and distribution, were found in western Europe and North America.

In the United States, municipalities provided regulations, mostly through charters that provided companies the right to construct lines in public streets and rights-of-way. These charters normally also were contracts whereby the gas company provided gas for street lighting at specified rates. Regulation of railroads by state boards or commissions was common in the eastern United States beginning around 1875. In 1885, Massachusetts became the first state to establish a Board of Gas Commissioners to regulate the gas business.[21]

However, charters were not always exclusive. Often companies were chartered to supply the same areas. Intense competition sprang up, with one company laying mains down one side of the street and another down the other side. Competition led to inefficiencies and economic trouble. Consolidation was the answer to the fight for territory and reckless pricing practices. In 1884, after nine months of discussion, the New York, Manhattan, Metropolitan, Municipal, Knickerbocker, and Harlem gas companies merged into the Consolidated Gas Company of New York.[22] Others followed suit over time. If consolidation was good for manufactured gas companies, adding the rapidly growing electric business to the mix was even better, and a number of combined manufactured gas and electric companies sprang up. Consolidation brought stability, but increased regulations were not far behind.

Meanwhile, back on the crude oil transportation side, battles of words, lawsuits, and sometimes more raged between crude oil producers, shippers, refiners, and the railroads as each sought advantage over the other. These battles fueled federal regulations and investigations, leading eventually to the Sherman Antitrust Act in 1890 and the subsequent breakup of the Standard Oil Trust in 1911. These actions did not directly affect the gas distribution business and actually had little effect on natural gas transmission at the time. Later in the second half of the 20th century, however, gas and oil pipeline regulations on the federal level began to merge.

Oil drives the need for long-distance transmission lines

Exactly who first discovered oil, and who started the oil industry, is up for debate, with a number of countries claiming the honor. However, finding oil and starting an industry are two quite different matters. The beginning of the oil industry is generally credited to "Colonel" Edwin Drake, who drilled an oil well near Titusville, Pennsylvania, in 1859. Others quickly followed Drake's example. The world's appetite for kerosene spurred the oil industry as kerosene quickly displaced whale oil and candles as the major source of home lighting. Carts, wagons, and barges (and in Azerbaijan, even peasants carrying oil-filled wineskins) were used to transport oil to railheads for movement on to refineries (fig. 2–6).

Fig. 2–6. Oil tank wagon and train
Courtesy: Buckeye Pipeline Company.

To the chagrin of oil promoters, gas was sometimes discovered as they drilled for oil. With no good way to transport the gas to market, it was usually flared or simply released into the atmosphere. The need to move large quantities of oil produced a new breed of engineers and entrepreneurs—pipeliners.

There were many small projects that demonstrated the potential for pipelines. One such project was constructed by Barrow and Company in 1863. They built an approximately 1,000-foot pipeline from oil wells to their refinery near Tarr Farm, Pennsylvania. However, leaks generally plagued the first pipelines due to poor quality pipe and joining techniques. The first technical and commercial success was a 2-inch diameter crude oil pipeline that spanned five miles from Pithole, Pennsylvania, to Miller's Farm on the Oil Creek Railroad. This line, built by Samuel Van Syckle in 1865, initially used three, and then four, pumps to move the oil.[23] Oil pipeliners continued to construct longer and larger lines as better pipe and joining techniques were discovered. Gas took a backseat to oil but benefited hugely from pipeline techniques discovered in the quest to move oil.

Natural gas transmission lines

In 1870, the Rochester Natural Gas Light Company constructed a 25-mile pipeline to supply natural gas from a well near Bloomfield, New York, to Rochester, New York. Unfortunately the project failed after two years as the pipeline, constructed of hollowed-out Canadian white pine logs, fastened together with bands of iron, developed excessive leaks. Consumers in Rochester also discovered that the natural gas burned hotter but produced less light than manufactured gas (fig. 2–7).

With the Rochester Natural Gas Light Company as a role model, producers began to market gas from gas wells. They even found a market for the *associated gas* that arrived at the surface mixed with the crude oil. Natural gas had to be separated from crude oil anyway at the wellhead before the crude oil was transportable and marketable. In the 1880s, Pittsburgh, by then a city with heavy industry, became the central focus for natural gas marketing and pipeline construction. Five hundred miles of pipelines brought natural gas from the oil and gas fields of western Pennsylvania, feeding 200,000 cubic feet per day of gas to many Pittsburgh businesses. The natural gas fueled 10 iron and steel mills, six glass-making factories, including the original Pittsburgh Plate Glass (now PPG) facility, and every brewery in town, including the legendary Pittsburgh Brewing Company, makers of Iron City beer. It even supplied the Simpson Natural Gas Crematory.

Fig. 2–7. Log pipeline and cast iron valve. This section of log pipe and the valve unearthed in the 1930s formed part of the 1870 pipeline to Rochester.
Courtesy: National Fuel Gas.

By 1886, the largest natural gas pipeline up to that time was constructed by the United Natural Gas Company. The 8-inch-diameter wrought iron pipeline stretched into Buffalo, New York, from Pennsylvania.[24] In 1880, the first compressor was used by the Bradford Gas Company. The 580-horse power duplex steam-driven compressor had a compressing capacity of 5 million cubic feet per day and a discharge pressure of 60 psi.[25] In 1887, the Dresser coupling was invented, which used a rubber ring to seal the connection between pipes. The coupling was later improved by adding a two-part mechanical device that pulled the coupling together. Dresser couplings saw widespread use into the 1920s, when they were replaced by welding.[26] By 1899, gas compressors had reached 1,000 horsepower (fig. 2–8).

While compressor and pipe technology had progressed considerably, much of pipelining still required a massive amount of manual labor. Digging ditches for the pipe with a pick and shovel took time and energy (fig. 2–9).

Fig. 2–8. Old Battleship compressor. This compressor was installed in Mt. Jewett, Pennsylvania, and was nicknamed the Old Battleship due to its size and shape.
Courtesy: National Fuel Gas.

Fig. 2–9. Digging ditch with pick and shovel
Courtesy: National Fuel Gas.

In any event, as the century turned, considerable knowledge existed about manufactured gas and its distribution, and there were volumes to prove it. A partial list includes the following:

- *A Practical Treatise on the Manufacture and Distribution of Coal-Gas*, by Samuel Clegg Jr., published about 1841.
- *Gas Works: Their Construction and Arrangement, and the Manufacture and Distribution of Coal Gas*, written by Samuel Hughes around 1855 and already in its seventh edition by 1885.
- *The Chemistry of Gas Manufacture*, by W. J. Atkinson Butterfield, published in 1886.
- *Handbook for Gas Engineers and Managers*, by Thomas Newbigging, in its fifth edition by 1889.
- *The Gas Engineer's Pocket-Book*, by Henry O'Conner, published in 1897 and in its second edition in 1901.

These works played critical roles in moving the industry forward.

Glory and Fall of Manufactured Gas: 1900 to 1950

Coal, and the gas manufactured from it, was king in western Europe and the United States as the new century began. It accounted for roughly 70% of US energy needs, while wood provided about 20%, and oil and natural gas together accounted for only 5% (fig. 2–10). At that point manufactured gas dominated the street lighting market (fig. 2–11).

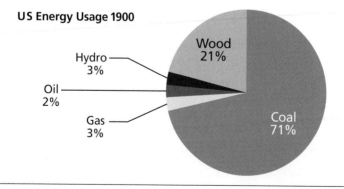

Fig. 2–10. US energy sources in 1900
Source: Compiled by Pipeline Knowledge & Development from statistics taken from the US Energy Information Administration Web site.

Fig. 2–11. Philadelphia gas street lighting, circa 1911
Courtesy: PhillyHistory.org, a project of the Philadelphia Department of Records.

As early as 1890, electricity was overtaking the street lighting market, and the cheaper natural gas was pressuring its way into the heating market. Coal use has continued into the 21st century, but manufactured gas has been replaced as a fuel source.

North America and parts of western Europe

By the first half of the 20th century, manufactured gas was in use worldwide, but from a practical standpoint, widespread usage was limited to the United Kingdom, France, Germany, Italy, the Netherlands, Belgium, Canada, and the United States. During the same time period, natural gas was used in limited ways in several countries, but significant use was limited to North America. Natural gas did not enter the European markets in any quantities until the 1960s. So, during the first half of the 20th century, discussing distribution pipelines meant discussing pipelines in the context of North America and about six European countries. Discussing natural gas transmission pipelines during the same time frame meant discussing pipelines located in North America.

Distribution

Technology. Cast iron pipe joined by the bell and spigot method was the preferred technology for mains at the turn of the century (fig. 2–12).

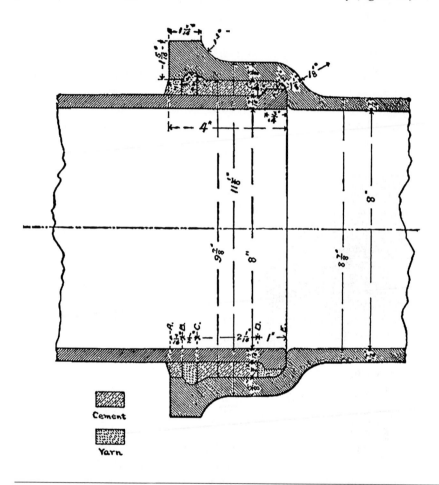

Fig. 2–12. Bell and spigot joint. In this case the space is filled with sealing yarn or jute, then cement, and then another layer of yarn, which is then sealed with cement. An earlier method for sealing used lead rather than the less-expensive cement.
Source: From *A Manual of Gas Distribution*, edited by Walton Forstall (1920).

Service connections were primarily wrought iron pipe with screwed connections. By 1950 rubber gaskets had replaced the sealing yarn and jute, and connections had been modified slightly to accommodate the new material.

Cast iron was less susceptible to rusting than wrought iron and steel. Consequently, it was preferred until the industry discovered that coating the outside of these pipes with tars or other coatings prevented, or at least reduced, external corrosion considerably. Cast iron remained the piping material of choice for mains until it was displaced by steel as steel making, pipe manufacturing, and welding methods improved.

Cast iron was rather brittle and subject to cracking. Since the pipes were not physically connected together until the invention of joint clamps, the spigot end had to be force fitted into the bell end, but the force could not be too great or the pipe would crack. When streets or underground obstructions were encountered, the pipes were sometimes installed above ground on bridges (fig. 2–13).

Fig. 2–13. Pipe bridge construction. Taken November 5, 1924, this picture shows construction of a pipe bridge crossing Broad Street at Norris Street in Philadelphia. Note the two workers on top of the bridge at the left and the assorted pieces of pipe at street level.
Courtesy: PhillyHistory.org, a project of the Philadelphia Department of Records.

As the gas moved from point of manufacture to point of consumption, water and other vapors introduced during the manufacturing process liquefied and dropped out of the stream. Mains and service lines were carefully designed and installed at slight inclines. Service lines, for example, sloped slightly to the main, causing any condensation to flow backward into the main. *Drips*, essentially small tanks, were installed at regular intervals along the line at the bottom of slopes to catch the condensation. These drips were then drained, sometimes as often as once per week (fig. 2–14).

Fig. 2–14. Horse-drawn drip wagon
Source: From *A Manual of Gas Distribution*, edited by Walton Forstall (1920).

Leaks were repaired temporarily by rubbing soap into the leak to seal it off and then wrapping the pipe with muslin cloth until the leaking pipe could be replaced or repaired more permanently. When a section of cast iron pipe was removed or when two pipes in the same ditch needed to be joined, the connection was often made by folding the joint in. Collars were installed on each pipe. Then two more pieces of pipe were precisely cut. The ends of these two pipes were put into the already installed collars, but at an angle that just allowed a third collar to be installed between the two pipe pieces. The pipes were then pushed straight, forcing the ends of each further into the collar. The collars were then caulked and sealed.

When repair work involved cutting into a pipe carrying gas, the work was sometimes conducted *live*, that is, without stopping the gas. As long as the gas did not light or asphyxiate the worker, everything was fine. Usually, though, the gas was *bagged off*, a normally straightforward procedure that involved installing two taps on the line, one on either side of the repair area. Deflated rubber or canvas bags were stuffed through the tap and then inflated, with the air forcing them against the pipe and shutting off the flow of gas. After completion of the repair, the workers deflated the bags and pulled them out, installing a plug to seal the tapped hole.

Organizations. Organizations varied depending on city size and other factors. In general, the leader of the distribution department, usually called the *distribution engineer*, reported to either an operations manager or the company president. The city was divided into geographic operating districts, each with a superintendent who managed the day-to-day operations and maintenance. Other superintendents managed functions supporting the districts (fig. 2–15).

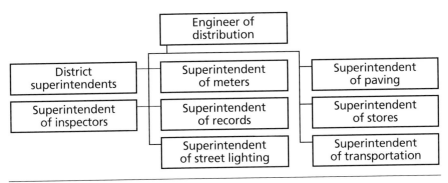

Fig. 2–15. Suggested distribution organization structure
Source: Prepared by Pipeline Knowledge & Development from information contained in *A Manual of Gas Distribution*, edited by Walton Forstall (1920).

The various titles give an overview of the activities needed to support distribution activities.

Business and regulations

As in the previous century, gas usage and its regulation were essentially limited to Europe and North America.

The United States. The sugar trust, meatpacking trust, tobacco trust, oil trust, and others grew during a time of industry consolidation in the late 19th and early 20th centuries in the United States. Power was no exception, as holding companies owning gas manufacturing, electrical generation, distribution of both gas and electricity, and natural gas transmission formed and gained monopoly and near-monopoly power in their geographic areas.

Prior to the advent of transmission lines, gas was produced and consumed locally, so logically local city or town regulations applied.

The newly formed power companies accepted the fact they would be regulated and thus lobbied for systematic state regulations, rather than patchwork local ordinances. New York and Wisconsin established public service and public utility commissions in 1907, charged with regulating the intrastate gas market. Other states quickly followed suit. Several states between 1911 and 1928 attempted to regulate interstate movement of gas, but the Supreme Court ruled such regulation unconstitutional.

During World War I, the US Fuel Administration (USFA) was created, marking the first time the US federal government became involved directly with the gas industry. The USFA attempted, not always with cooperation, to direct and reallocate gas usage in support of the war effort. As the war ended in 1918, the gas and power industry in the United States was characterized by a series of large regional holding companies that dominated the power business.

Consolidation continued, and the power companies grew, prompting governmental concern. In 1928, the US Senate asked the Federal Trade Commission (FTC) to study and report on the state of public utility holding companies. The report, released in 1935, contained 96 volumes, and led to the Public Utility Holding Act (PUHA), which required all public utility holding companies to register with the Security and Exchange Commission and divulge the following:

- Copies of the charter or articles of incorporation plus any partnership agreements, bylaws, trust indentures, mortgages, underwriting arrangements, and voting trust arrangements
- Full revaluation of the financial structure of the company
- A list of all officers and directors
- Explanations of any bonus and profit-sharing arrangements
- Provisions of any contracts for materials, services, or construction
- Consolidated balance sheets and comparable information

The PUHA also limited utility companies to single, locally managed systems and separated natural gas and electric operations.

Around the same time, several other pieces of US federal legislation important to the gas industry passed. These included the following:

- Federal Power Act
- Natural Gas Act
- Securities Act
- Securities and Exchange Act

Not surprisingly, one of the primary lobbying groups for all this regulation was the coal industry, which saw a large portion of their market being eroded by natural gas.

During World War II, the Petroleum Administration for War (PAW) was created. However, oil, and not gas, was of primary concern to the PAW as they sought to ensure sufficient supplies of oil to fuel the naval fleet and army tanks and trucks. The PAW did form a subgroup, the Natural Gas and Natural Gasoline Division, which was responsible for overseeing natural gas and the liquids extracted from it. Most of the manufacturing supporting the war effort was located in the northeastern United States, so during the war, extensive drilling for natural gas in the Appalachians was encouraged.

Two key events rising from World War II were construction of the Big Inch Pipeline and the Little Big Inch Pipeline. The Big Inch Pipeline was a 22-inch diameter crude oil line from Longview, Texas, to Phoenixville, Pennsylvania, while the Little Big Inch Pipeline was a 20-inch refined products pipeline from Beaumont, Texas, to New Jersey and Pennsylvania. Following the war, both of these lines were sold to Texas Eastern Transmission and converted to natural gas service. Gas from the Gulf Coast met northeastern US demand, as the drilling in the Appalachians had nearly exhausted that gas supply.

As mentioned previously, natural gas and manufactured gas are actually two different types of gases. Natural gas is methane, with a British thermal unit (Btu) content of around 1,050, while manufactured gas is a mixture of other gases, with a Btu content between 450 and 600. Sometimes the two gases were mixed in the same system, but not always with dependable results.

In the 1940s, conversion from manufactured to natural gas was underway in many US cities. The process involved sectioning the mains and isolating an area. On "conversion day" the gas company brought in scores of workers. The manufactured gas was shut off and flared. Then all appliances were retrofitted with smaller orifice plates to accommodate the higher Btu content of the natural gas. Finally natural gas was introduced into the mains, pushing any remaining manufactured gas out through flares. The flare flame changing color was the signal the main was now full of the new gas. Finally, all appliances were relit, completing the conversion.

Conversion provided a benefit. The higher Btu content meant the system capacity suddenly nearly doubled, since not as much of the gas was needed to provide the same heat. However, it also presented a challenge. Natural gas, with its much lower liquids content, resulted in less maintenance related to the draining of the liquid in the required drips for manufactured gas lines. However, the liquids in the manufactured gas kept the yarn and jute moist and flexible so that they could seal off the bell and spigot joints. Dry natural gas meant these sealing components dried out and began leaking. Natural gas distribution workers were kept busy for the next several decades chasing leaks and resealing joints.

Between 1945 and 1954, natural gas doubled its US market share, moving up to 22% of the energy market.

The United Kingdom. According to the *National Gas Archive*, the Gas Regulation Act of 1920 established gas pricing "and introduced a national basis for testing and reporting gas quality."[27] In the 1930s, gas holding companies came into prominence in the United Kingdom. However, the extensive bombing and rocket attacks of World War II on the United Kingdom in general, and London in particular, significantly damaged the manufactured gas infrastructure. According to the *Archive*, "In 1944 the Ministry of Fuel and Power set up a Committee of Enquiry under the Chairmanship of Geoffrey Heyworth." Heyworth's report "formed the basis of the Gas Act of 1948, which nationalised the gas industry in England, Scotland and Wales."[28] At that time, there were 1,064 local gas undertakings divided into 12 autonomous area gas boards. The 12 presidents of these boards comprised the Gas Council.[29]

Transmission lines

By 1902, there were approximately 25,000 miles of natural gas transmission lines located in the United States. Of these, 68% were located in Pennsylvania, Ohio, West Virginia, and New York.[30] These lines operated at considerably higher pressures than distribution lines and were constructed of steel or wrought iron pipe. They used screwed connections. One end of the pipe had external threads, the other internal. One was inserted into the other, and the whole thing was screwed together with large tongs (fig. 2–16).

Over time, joining techniques turned to welding as first oxyacetylene, then acetylene, and finally electric arc welding took over (fig. 2–17).

Fig. 2–16. The tong gang. One pipe was held in place while the other was turned to screw them together.
Courtesy: Buckeye Pipeline Company.

Fig. 2–17. Acetylene welding
Courtesy: Anonymous contributor.

The first half of the 20th century involved learning how to build, start, and run larger compressor stations powered directly by gas engines rather than by steam engines. One early problem involved getting the huge gas engines started turning in the first place. Reportedly explosive charges were placed in the cylinders and detonated to get the engines going. During the same time, steel and pipe manufacturing processes improved, machines replaced hand ditching, motorized equipment replaced mules and horses, and telegraph lines gave way to telephone lines.

Pipeliners borrowed inventions from others to improve their work, or simply invented the equipment as needed. The town of Morrowville, Kansas, contains a tribute to the bulldozer, invented by local farmer James Cunningham with the assistance of J. Earl McLeod, a local draftsman. Watching the Sinclair Oil Company laying a line across Washington County, Kansas, Cunningham noticed ditching had been mechanized but backfilling was still performed with mules and dirt slips. He and McCleod used the frame of a Model T, installed assorted parts from a junkyard, and invented the bulldozer. They used the machine to backfill the line from Deshler, Nebraska, to its terminus in Freeman, Missouri.

The first long-distance, all-welded steel natural gas pipeline was laid by Magnolia Gas of Dallas in 1925. It extended from northern Louisiana to Beaumont, Texas, and was 217 miles long. In 1931, the first 1,000-mile gas line, 24 inches in diameter, was laid by Natural Gas Pipeline from the panhandle of Texas to Chicago.

Improvements to compression, materials, equipment, and joining techniques would continue and were shared between oil and gas transmission pipelines and gas distribution pipelines. By the middle of the century, natural gas transmission was well understood, and the technologies were in place to drive it forward.

Storage

Entering the century, gas lifters provided the needed storage to bridge cyclic demand with steady production (fig. 2–18).

Over time underground storage, aquifers, caverns, and pressure tanks took the place of lifters.

The first documented use of underground storage was in Welland County, Ontario, Canada, in 1915. In 1936, the United States had 13 storage reservoirs with a total capacity of 39 billion cubic feet of storage. By 1950, those numbers had grown to 125 underground storage fields and total underground capacity of 774 billion cubic feet.[31]

Fig. 2–18. Gas lifter. Concentric rings, with a water or oil seal between them, were nested inside each other. They expanded or contracted much like an accordion as gas entered or left the storage container. Capacities were on the order of 10 million cubic feet, but one historian remembers one as large as 17 million cubic feet.
Courtesy: National Fuel Gas.

Liquefied natural gas (LNG) allows natural gas to be stored as a liquid rather a gas, and it was developed based on compressor refrigeration technology first discovered in the 19th century. The first LNG plant was brought into service during 1912 in West Virginia. Additional plants were constructed to store gas during periods of low use so it could be injected into the system when needed. Tragically, in 1944 an explosion of gas leaking from an LNG storage tank in Cleveland killed 130 people and injured many more, calling the safety of LNG into question. As more gas reservoirs became depleted and available for storage, and as technology improved to allow higher transmission pressures, use of LNG for storage became less economic.

Turbulent Times to Shining Star: 1950 to 2000

By the midpoint of the 20th century, well-developed gas distribution systems were running in the United States and in several countries in western Europe. In the United States, manufactured gas was still being used, but natural gas was poised to take over. The conversion to natural gas was completed before the 21st century began.

In 1951, the Lacq gas field in southwest France was discovered by Elf Aquitaine. The first transmission system in France, 4,000 kilometers long, was built to supply gas to southwest France, Brittany, and the Paris area. Cities switched progressively to natural gas, and by 1965 approximately half of France was supplied with natural gas.

In 1959, a converted freighter, named the *Methane Pioneer*, moved LNG from Louisiana to Britain, proving the technical viability of LNG transportation by ship. About five years later, Britain turned to a liquefaction plant in Algeria for LNG imports until discovery of North Sea natural gas in the 1970s. By 1977, all appliances in the United Kingdom were converted to natural gas. Japan began importing LNG in 1969, and by 1988, they were completely converted to natural gas. Similar conversions occurred elsewhere.

Natural gas, the ugly stepchild to oil up until about the 1900s, has become the darling of business people and environmentalists alike. However, the regulatory road has been bumpy and is likely to remain so.

Distribution

Conversion to natural gas and the vexing problem of how to seal leaking joints, along with how to efficiently install and maintain distribution mains and service lines, drove some ingenious solutions. One involved pumping a tar-like substance into the line to seal up leaks. Others involved mechanical clamps to hold the joints together, encapsulating the outside of the joint with an epoxy solution, and even inserting remote-controlled devices into the line to caulk joints from the inside. Complete liners of copper, plastic, composite materials, and other materials to contain the pressure were also tried with varying degrees of success. Experimentation and trial and error proved the good ideas and disposed of the bad ones.

Plastic pipe. The advent of plastic pipe, first discovered in the 1930s but not put into large-scale use until the mid-1950s, significantly changed distribution industry practices. After several false starts, the industry settled on polyethylene pipe. Now more than 90% of all new distribution mains and service lines installed around the globe are constructed of high-density polyethylene pipe.

Trenchless installations. Boring, ground piercing, and finally horizontal directional drilling (HDD) have reduced the cost and improved the safety of installing mains and service pipes without trenching. Service connections and disconnections, along with some forms of maintenance, were much improved with keyhole and small-hole techniques.

Communications, computers, and control. Formerly, operators read gauges and meters and then telegraphed, and later phoned, the information into a central dispatch center. Centralized dispatchers responded with orders, and operations were performed manually. Now, though, communications, sensors, transmitters, and actuators handle many of these tasks automatically and routinely. Smart meters are even being installed at residences, reducing the need for meter readers.

Storage. Depleted gas reservoirs still provide the majority of storage, but salt caverns are being mined to provide high-deliverability storage. LNG storage is coming back into use, particularly at import terminals.

LNG. LNG is simply natural gas in a liquid rather than a gaseous state. Liquefied natural gas requires about 600 times less storage space than natural gas at standard temperature and pressure. When pipeline transport is not economic, LNG transport by ship may be. Japan and South Korea, for example, depend primarily on the import of LNG to meet their natural gas needs. Once returned to the gaseous state, natural gas can be injected into distribution systems just like natural gas that has never been liquefied. In addition to water transport, LNG is showing signs of coming back for storage to load balance across demand changes.

Business and regulations

The 50 years between 1950 and 2000 saw vast changes to natural gas regulations and the natural gas transmission and distribution businesses. Covering these changes is beyond the scope of this chapter but they are covered in more detail later in this text.

Business. Generally the natural gas distribution business exited the 20th century as a freer industry, with more private ownership, than when it entered the second half of the century. In the United States, the business model had been producers selling to transmission pipelines, who in turn sold to local distribution companies and some large end users. Very few transmission pipeline hubs or interconnects existed, and LDCs purchased gas from the limited number of transmission pipelines serving their area. Beginning in the 1970s, however, the Federal Energy Regulatory Commission (FERC) changed the business model. LDCs can now purchase from anyone. The transmission pipelines serve essentially as common carriers and are not involved now in buying or selling gas. They only provide a transportation service. Europe is trying to move toward a similar model to widen supply alternatives and improve competition.

LDCs are owned in some cases by investors, but in other cases they are owned by local municipalities. In Russia, China, Ukraine, and many other socialist and former socialist countries, the local distribution function is still owned by an arm of the government.

Distribution economic regulations. The economic model in play, free enterprise versus centrally planned, or somewhere in-between, is a major factor in determining economic regulations. Over time, it became apparent that a single infrastructure was a more efficient economic model than several competing infrastructures. This resulted in distribution systems that were monopolies.

Free enterprise economics have largely migrated to allowing LDCs to recover their costs of operations and in addition earn a return on invested capital. The factors, calculations, and exactly what costs are allowed and how they are treated may vary, but the concept remains the same.

Centrally planned economies establish rates to promote their ends. For example, China, where less than 4% of the energy use is from natural gas, establishes rates based on regional pipeline infrastructure development. Rates for the industrial customer class generally remained well below market rates. While the priorities continue to evolve, China has historically favored manufacturing and fertilizer gas users, with these sectors paying lower prices than residential customers.[32]

US transmission economic regulations. The United States went through a total regulatory reorganization of the natural gas transmission industry during the last half of the 20th century. This reorganization affected the LDC business model and is being studied by other countries as they attempt to create a more competitive and open market.

Following World War II, the Federal Power Commission was struggling with how to regulate natural gas transmission pipelines. They looked to the Interstate Commerce Commission, the entity regulating railroads, and to local utility regulators, which regulated local distribution pipelines, for clues. Based largely on these models, they adopted a three-pronged approach to regulations:

- Control over entry and competition through a certification process
- Regulation to assure "reasonable" rates based on a fair rate of return on a fair level of investment
- The imposition of standard accounting practices to simplify both financing and rate setting

The ensuing years marked a confusing and tumultuous time. From price controls, to limiting and then encouraging competition, the Federal Power Commission experimented with how they could best apply these principles to the natural gas pipeline industry. Court cases abounded, including some at the Supreme Court level. Artificially low pricing drove demand but limited new natural gas exploration. Regulations caused gas shortages in some regions as the highly regulated commodity sought its highest value markets.

Finally in 1978 the Natural Gas Policy Act, passed as part of the National Energy Act, abolished the Federal Power Commission, replacing it with FERC. Both the Natural Gas Act and the Natural Gas Policy Act maintained the natural gas pipeline business model as a "bundled model." Gas pipelines purchased gas from producers, transported it to the marketplace, and sold it to customers. Customers could not buy transportation services separately from the gas itself. FERC Order No. 436, issued in 1985, began to change the model by establishing a voluntary process whereby interstate pipelines could sell transportation services separately, setting them up as merely gas transporters. FERC Order No. 636 "unbundled" transportation, storage, purchasing, and marketing of natural gas, making what was formerly voluntary—transportation for others—mandatory.[33]

Safety and land use regulations. The last half of the century saw an explosion of safety and land use regulations. Safe operations are in the industry's best interest, so industry members share experiences and knowledge, which are formalized into industry standards. Three of the oldest industry standards-setting organizations are the British Standards Institute, established in 1901 as the Engineering Standards

Committee; the American Gas Association, founded in 1918; and the American Petroleum Institute, which began in 1924. These organizations have overseen development of hundreds, if not thousands, of standards covering areas such as materials, construction, maintenance, and operations.

In spite of standards and best intentions, accidents happen. However, no accident is ever acceptable. Additionally, everyone wants conveniences like hot water, but no one wants the imposition of a transmission line in their backyard. Chronicling the growth of safety and land use regulations affecting LDCs is beyond the scope of this text. Suffice it to say, politicians and regulators do their best to react to public sentiment but find themselves at a disadvantage since the LDCs (in most cases) know more about operations and safety than do the regulators. The regulators want to work with the industry to develop the best safety and land use regulations, but if they work too closely, they might be accused of cooperating with the companies. Various accidents have influenced the development of regulations, such as the 1937 explosion of a school in New London, Texas, which killed about 300 people. It resulted in the requirement of the addition of odorants to natural gas. Another such accident occurred with the rupture of the Tennessee Gas Pipeline in Natchitoches, Louisiana, in 1965, with 17 fatalities. This accident resulted in the enactment of the 1968 Natural Gas Pipeline Safety Act in the United States. These are only two examples of catastrophes that have significantly impacted safety and land use regulations.

Summary

- Natural gas has been known for thousands of years.
- Manufactured gas was developed around 1800.
- The first gas was used for lighting factories and streets.
- Gas regulations started at nearly the same time as the industry, owing to the need for city charters to lay pipes in city streets.
- Improvements in materials and technology were needed to allow the widespread use of gas.
- Gas, kerosene, and electricity fought for the home and street lighting markets.
- Manufactured gas was in widespread use in Europe and North America by 1900.

- Gas transmission technology was aided by the developing oil pipeline industry.
- World Wars I and II were instrumental in moving forward natural gas usage in the United States.
- Conversion from manufactured gas to higher energy natural gas provided additional capacity in existing lines.
- Worldwide conversion to natural gas occurred from the 1960s through the 1990s.
- LNG has been well-known and extensively used since the 1970s.
- Plastic pipe and fittings reduced costs and improved efficiencies beginning in the 1970s.
- Technology changes were key enablers of the distribution industry, which often works in congested urban environments.
- Wholesale changes to the natural gas business model in the United States—particularly transmission—occurred from about 1970 to 2000.
- Other countries are looking at the US model as they attempt to develop their own unique brand of open markets and improved competition.
- Regulators face challenges as they attempt to promulgate wise regulations without being seen as too cooperative with industry.
- Natural gas is the current best environmental choice but has only been used in a serious way in the world since about 1950.

Notes

1. John Roach, "Delphic Oracle's Lips May Have Been Loosened by Gas Vapors," *National Geographic News* (July 31, 2001), http://news.nationalgeographic.com/news/2001/08/0814_delphioracle.html.
2. Oscar E. Norman, *The Romance of the Gas Industry* (Chicago: A.C. McClurg & Company, 1922), 20.
3. Ibid., 35.
4. Ibid., 22.
5. National Grid, "The Pioneers of the Gas Industry," *National Gas Archive* (National Grid, 2005), http://www.gasarchive.org/Pioneers.htm.
6. London County Council, "Pall Mall, South Side, Past Buildings: Nos. 93–95 Pall Mall: F.A. Winsor and the Development of Gas Lighting," *Survey of London: Volumes 29 and 30: St James Westminster, Part 1* (London: London County

Council, 1960), 352–354, British History Online, http://www.british-history.ac.uk/survey-london/vols29-30/pt1/pp352-354.
7. Allen W. Hatheway, *Remediation of Former Manufactured Gas Plants and Other Coal-Tar Sites* (Boca Raton, FL: CRC Press, 2008).
8. Daniel W. Mattausch, "David Melville and the First American Gas Light Patents," *The Rushlight* (The Rushlight Club, December 1998), http://www.rushlight.org/articles/gas.html.
9. Christopher J. Castaneda, *Invisible Fuel, Manufactured and Natural Gas in America, 1800–2000* (New York, Twayne Publishers, 1999), 22.
10. Samuel Clegg, *A Practical Treatise on the Manufacture and Distribution of Coal Gas*, 5th ed. (New York) D. Van Nostrand, 1868), 17.
11. Cast Iron Soil Pipe Institute, *Cast Iron Soil Pipe and Fittings Handbook* (Chattanooga, TN: Cast Iron Soil Pipe Institute, 2006), 1, http://www.cispi.org/CISPI/files/8b/8b6004bc-1eaa-405c-b452-29061680a58e.pdf.
12. Walton Forstall, ed., *Manual of Gas Distribution* (Philadelphia: U.G.I. Contracting Company, 1920), 407.
13. Clegg, *A Practical Treatise*, 21.
14. Norman, *The Romance of the Gas Industry*, 47–48.
15. Paul H. Giddens, *The Birth of the Oil Industry* (New York: The Macmillan Company, 1938), 19.
16. Norman, *The Romance of the Gas Industry*, 125.
17. Ibid., 128.
18. Castaneda, *Invisible Fuel*, 44.
19. Ibid.
20. Jocelyn Hunt, *Britain, 1846–1919* (London and New York, Routledge, 2003), 64.
21. Louis Stotz, *History of the Gas Industry*, in collaboration with Alexander Jamison (New York: Stettiner Bros., 1938), 447.
22. Castaneda, *Invisible Fuel*, 64.
23. Giddens, *The Birth of the Oil Industry*, 143.
24. David A. Waples, *The Natural Gas Industry in Appalachia* (Jefferson, NC: McFarland & Company, 2005), 138.
25. Ibid., 141.
26. Castaneda, *Invisible Fuel*, 46.
27. National Grid, "From the First World War," *The National Gas Archive* (National Grid, 2005), http://www.gasarchive.org/FirstWW.htm.
28. Ibid.
29. National Grid, "Nationalisation," *The National Gas Archive* (National Grid, 2005), http://www.gasarchive.org/Nationalisation.htm.
30. Waples, *The Natural Gas Industry in Appalachia*, 140.
31. T.C. Buschbach and D.C. Bond, *Underground Storage of Natural Gas in Illinois—1973* (Urbana, IL: State of Illinois, Department of Registration and Education, 1974), 3.

32. US Energy Information Administration, "China," *Country Analysis Briefs* (July 2009), http://www.eia.doe.gov/cabs/China/Full.html.
33. Thomas O. Miesner, *A Practical Guide to U.S. Natural Gas Transmission Pipeline Economics*, An Oil and Gas Research Center Report (2009), 13.

chapter **3**

How Pipelines Work

Opportunity is missed by most people because it is dressed in overalls and looks like work.

—*Thomas Edison*

Natural gas arrives on a nearly continual basis at water heaters, stoves, ovens, and furnace burner tips, and, just like with water to the faucet, few people consider how it gets there. Both water and gas pipelines seem simple enough—put it in one end, take it out of the other. The principles dictating the behavior of fluids (gas and water, for example) moving through pipelines are rather intuitive, but the calculations involved can be fairly complex. Thankfully this chapter uses only a few simple equations as it covers the properties of natural gas and how it flows.

The Physics of Fluid Flow

Though it may seem obvious, natural gas contained in pipelines and storage vessels conforms to the shape of those vessels and completely fills them up, an important characteristic of all gases. (This differs from liquids, which do not fill the container completely and have one free surface.) However, what happens when more gas is pushed into the line? Each molecule simply shoves its neighbor over to make room. This shoving increases the pressure inside the pipeline or storage vessel. If the pipeline is opened to let some gas out without putting any more in, the pressure inside the pipeline decreases.

Compressors shove molecules into gas pipelines, adding energy in the form of pressure. Valves let the molecules out at controlled locations and rates, thereby reducing the pressure inside the pipeline. The pressure in a pipeline that is shut down (nonflowing) along a level route is the same along its entire length. Molecules have weight, and gravity pulls on them. Thus if a pipeline is shut down and there are elevation changes along the route, the pressure is higher in the valleys and lower at the hilltops. However, once the pipeline resumes flowing, the pressure is nearly always lower as the fluid moves along. Natural gas is relatively light, so elevation change does not affect the pressure as much as for heavier molecules such as water. Nevertheless, engineers must still consider elevation changes as they calculate pressures.

Fluids always move from a point of higher energy to one of lower energy unless something like a closed valve stops them. Recall that in pipelines, energy is normally measured as pressure. As energy is added to pipelines by compressors, pressure builds. If the pipeline is opened at any point, flow starts. Even though fluids look slippery, friction occurs between the inside wall of the pipe and the moving fluid and between the molecules of the fluid. This resistance to flow must be overcome with energy. Friction generates heat, which is another way of saying it converts pressure energy to heat energy. This heat energy is transferred to the fluid and dissipated into the surrounding environment. Figure 3–1 shows a simple example of a level pipeline. The compressor station at the origin adds pressure that is then consumed by friction and lost as a motive force.

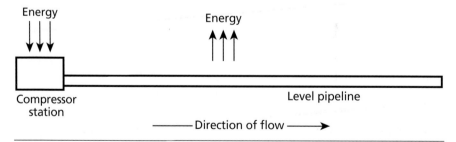

Fig. 3–1. Fluid flow along a level line. Pressure all along the line is the same when the line is not flowing.

The equation for the pressure at any point along this level line is as follows:

Pressure at any point = Pressure at origin
                              − Pressure loss due to friction

Add in the element of elevation and the equation becomes as follows:

Pressure at any point = Pressure at origin
                              − Pressure loss due to friction
                              ± Pressure due to elevation differences

Figure 3–2 shows a line with elevation changes.

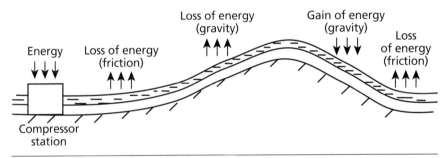

Fig. 3–2. Fluid flow along a line with elevation changes. Pressure at the top of the hill is lower than pressure at the bottom due to the effects of gravity, as long as the pipeline is not flowing.

Municipal Water Systems

Municipal water systems are a good analogy for natural gas distribution pipelines and illustrate the forces of gravity and pressure. The water tower, a huge tank on a hill or on a standpipe, is a familiar sight in most towns. Water moves from the tank, through the mains, and around town by energy from gravity, sometimes assisted by pumps located in carefully selected locations. Of course, the built-up energy of the water in the tank came from the pump that initially pushed the water up there.

At any faucet, the pressure from the elevated tank or pump is higher inside the faucet than the atmospheric pressure outside the faucet. When the faucet opens, water gushes out as the fluid seeks the lower pressure in the sink. The same thing happens with natural gas local distribution lines, but the valves are normally opened automatically by an appliance, such as the water heater, furnace, or gas dryer.

A water system is sometimes called a *hydrant system*. Water can come in from several receiving inputs and can leave the system from several thousand output points. That is okay because water is water, and the source of the water does not make any difference to the consumer. Natural gas distribution pipelines normally operate as hydrant systems. Natural gas enters and leaves the systems at multiple points. It is all the same and consumers do not care where it originates.

There is one important difference between water systems and natural gas distribution systems. Water is a liquid, but natural gas is a gas, and thus compressibility enters the discussion. Natural gas is much lighter than water, so elevation differences provide (or require) less energy. But natural gas is compressible, allowing the system to store pressure provided by the natural gas transmission pipeline when it delivered the gas to the local distribution lines. This stored pressure, and not the pressure caused by elevation, is the main force moving natural gas through the distribution lines to homes, schools, hospitals, and businesses.

Friction Losses, Pipe Lengths, and Flow Rates

Pushing the domestic water system analogy along, consider the lowly garden hose. A 50-foot length of hose attached to a spigot delivers a certain amount of water through the nozzle. A 100-foot length delivers less, and the reason is friction loss. As water flows along, water molecules rub against the walls of the hose and each other, generating friction and dissipating energy (losing pressure). The longer the hose, the greater the friction, the more the pressure drop, and the less the flow.

Most people have a spray nozzle (valve) at the end of their garden hose. It allows them to stop the flow without turning off the spigot. With the nozzle shut, the pressure at the spigot is the same as the pressure at the nozzle. Since the hose is not flowing, all the pressure in the hose is potential energy manifested as pressure (versus kinetic energy, the energy of motion). There is no friction loss since there is no flow. Each molecule is pushing against both its upstream and downstream neighbor equally hard, and the one at the very end is pushing against the spray nozzle. They bounce around inside the hose but none moves very far. When the nozzle opens, the molecule right next to the nozzle suddenly has only the air (atmospheric pressure) outside the hose to push against it. It leaps out, setting up a chain reaction and allowing flow. Since there is now flow, there is also friction loss.

As the nozzle opened, a number of things happened:
- The diameter of the plastic garden hose had stretched slightly as the pressure built when the nozzle was shut. When the nozzle opened, the pressure dropped, the hose contracted, and with less space in the hose, extra water surged out of the nozzle.
- A pressure wave moved quickly from the nozzle back through the hose. Molecules in the hose went successively from stationary to moving in a short time, demanding energy and causing friction.
- Eventually (after a few tenths of a second), the flow settled down to a constant rate.

As the nozzle closed, the following happened:
- The flow decreased and the friction losses decreased.
- A pressure wave moved from the nozzle back. If the nozzle was closed quickly, the pressure wave could have been large enough to make the hose surge ("jump").
- The pressure equalized (increased) in the hose, which expanded.

In this simple garden water hose example, any pressure wave moves quickly, since water is a liquid and essentially not compressible. The compressible nature of natural gas, however, means pressure waves are dampened as they move through the "spongy" gas.

The air compressor and air tank are useful in demonstrating another property of compressible fluids like natural gas. An empty air tank in a garage is the same temperature as the garage. Turn on the air compressor and pump up the air tank. There is now more air in the tank, and the tank feels hotter. The compressor forced air molecules closer together, meaning they cannot move around as much. The energy that previously caused them to bump around had to go somewhere. It turned into heat, which is an important concept in natural gas compressor operations. Conversely, reducing the pressure absorbs heat (fig. 3–3).

Another familiar example, the balloon, demonstrates the power of compressibility. Fill a balloon with water, hold the end closed, lay it on a table, and let go of the end. The water gushes out. Fill an identical balloon with air, lay it on the same table, and release the end. The balloon flies off the table and all around the room.

Fig. 3–3. Dropping pressure across a control valve. The pressure drop across this valve absorbs heat, cooling the pipe. Note the frost formed on the valve as the humidity in the air condensed and froze.

To return to the garden hose analogy, a nozzle usually has a number of settings, allowing different amounts of water to escape in different patterns. (Of course, some people expertly use their thumbs instead.) When the chosen nozzle pattern allows more water to come out, the flow rate through the hose, and consequently the pressure loss as the water moves to the nozzle, is higher. That, together with the much larger Bernoulli principle (discussed later in this chapter) will cause the water not to shoot out very far (fig. 3–4).

Smaller patterns allow less water to escape, slowing flow and reducing pressure loss. The flow, which is measured by the amount, not the speed, of water passing a certain point in a certain time, is slower with smaller openings. Counterintuitively, the water shooting out is traveling faster (at a higher velocity). Thus less water comes out, but it comes out faster (fig. 3–5).

Why does this occur? Less flow and less pressure loss through the hose mean more pressure remains at the end of the hose. But immediately outside the hose, the pressure must be equal to atmospheric pressure. Thus all the residual pressure is turned into kinetic energy, the energy of motion, across the nozzle, and the water shoots out faster and farther.

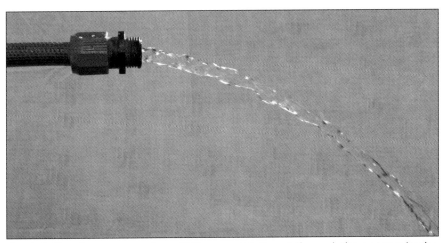

Fig. 3–4. Flow from an open hose. With no restriction at the end, the water exits the hose at high flow rates but low velocity.

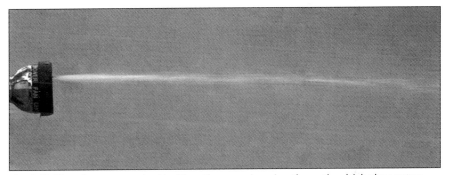

Fig. 3–5. Flow from a nozzle. Limiting the opening size through which the water exits reduces the amount of water exiting, but it exits at higher velocity.

The following summarize the fluid flow discussion:
- Friction causes resistance to flow.
- Faster flow rates produce more friction than slower rates.
- Changing pressures can cause the conduit to expand or contract.
- The size and pattern of the opening determine the pressure loss and therefore velocity through the opening.
- As the flow starts or stops, the pressure is different than at steady state.
- Changing one variable, pressure or flow rate, can change the other, and changing the length changes both.

Splitting flows

Moving inside the house offers another analogy. When someone takes a shower and someone else in the house unconscionably flushes a toilet, some of the water previously going to the shower goes to the toilet. A sudden reduction in water flow to the shower occurs. To the chagrin of the showerer, sometimes the cold water flow drops more than the hot water flow, and the shower changes momentarily to scalding (fig. 3–6).

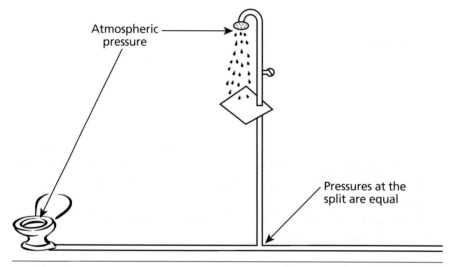

Fig. 3–6. Shower and toilet example. Pressures must be equal at the split.

When valves are open in more than one location, various laws of physics redistribute flows in ways that balance pressures and flow rates. Where the water flows split, the following equations apply:

Pressure at split = pressure at shower
+ pressure loss between shower and split

Pressure at split = pressure at toilet
+ pressure loss between toilet and split

That is the basic lesson in the physics of flow, but it only covers half the fun of hydraulics. The second part consists of the laws of fluid mechanics, most of which were developed by the great minds of the 17th, 18th, and 19th centuries. But before that comes the properties of fluids. Water has been the example for a lot so far, and natural gas follows the same laws of physics. However, differences in properties between water (a liquid) and natural gas (a gas) dictate the dissimilarity of their flow behavior.

Hydraulic Properties of Hydrocarbon Fluids

Properties of the fluids transported in pipelines are determined by the molecules that comprise the fluid and how those molecules react with each other and the environment. Natural gas pipelines carry mainly methane molecules, so the following important hydraulic properties are very similar between lines:

- Density
- Viscosity
- Specific heat
- Vapor pressure
- Compressibility

Even so, pipeline designers make sure they understand the properties of the specific streams intended for their systems as they design the lines. Operators closely monitor properties to ensure they remain within acceptable ranges.

To make life unnecessarily complicated, the measurers measure these measures by more than one set of standards.

Measures and dimensions

Two basic ways of measuring have evolved over the years: the metric system (now commonly referred to as the *International System of Units*, abbreviated simply SI) and the English or US customary system (USCS). Both are used in hydraulics. Length and time are easy enough to understand. In the USCS, lengths come in inches, feet, yards, and miles. The more logical SI system uses the meter as the basic measure, and divides it into centimeters, millimeters, and even smaller divisibles of 10, as well as the larger kilometers. Time is blessedly expressed in both systems in the familiar seconds, minutes, and hours.

The third important set of dimensions in hydraulics, force and mass, is a bit more difficult. People easily confuse weight and mass, or force and pressure, which does not usually matter unless they happen to be working on pipeline hydraulics. Suffice it to say the engineers must keep all the units and nomenclature constant and consistent, but for purposes of this book, pressure is expressed in the USCS as pounds per square inch (psi) or the smaller ounce per square inch, called here simply *ounces*. In the SI system, pressure is measured as *pascals* (Pa, or the larger kilopascal, kPa), named for the French mathematician, physicist, and religious philosopher, Blaise Pascal (1623–1662). (One pascal equals one newton per meter squared.) Another common pressure measure is a *bar*, equivalent to 100 kilopascals. Strictly speaking, the bar is neither a USCS nor an SI measure but is widely accepted for use in the SI system. Bars are handy because 1 bar is approximately equal to atmospheric pressure; 1 bar equals about 14.5 psi.

Pressure. As mentioned in the previous section, pressures themselves are measured in pounds per square inch (psi) in the USCS and in pascals (Pa) or kilopascals (kPa) in the SI system. This book uses pounds per square inch. Conversion between the two is rather simple, however: 1 psi equals 6,895 Pa and 1,000 psi equals 6,895 kPa.

Pressure can be measured as absolute pressure (psia) or gauge pressure (psig). Most pressure gauges are calibrated to read 0 at atmospheric pressure. When a tire gauge shows 30 psi, the pressure inside the tire is actually about 44.7 psi. Thus, the following equation applies:

$$\text{Absolute pressure} = \text{Gauge pressure} + \text{Atmospheric pressure}$$

The standard atmospheric pressure is 14.7 psi at sea level, but ambient pressures change with the weather, and pressures differ from valleys to mountaintops. Knowing either atmospheric or gauge pressure is fine, but the user must know which they are using and adjust the number if needed. Most gas pipeline hydraulic equations use absolute pressure.

Temperature. Most people are familiar with temperature expressed in Fahrenheit or Celsius, but just like with pressure, absolute temperature is normally used for gas pipeline hydraulics. William Thomson (1824–1907), who was given the title Baron Kelvin after the river Kelvin that flowed by his university in Glasgow, Scotland, proposed an absolute temperature scale for the SI system around 1854. About the same time, some say 1859, another Scotsman, William John Macquorn Rankine, proposed an absolute scale for the USCS.

Both start at absolute zero, the temperature at which all molecular motion stops (–273.15° C and –459.67°F). Kelvin named the SI temperature scale for himself and set it equal to the temperature in Celsius plus 273.15. Somehow the *e* was dropped from Rankine's last name and the other scale became known as the Rankin scale. *Degrees Rankin* is calculated by adding 459.67 to the Fahrenheit temperature. Interestingly, *degrees Rankin* are abbreviated °R, but Kelvin omits the degree sign and is simply K. So the temperature at which water freezes is 0°C, 32°F, 491.67°R, and 273.15 K.

Properties

Density. Strictly speaking, *density* is the quantity of a fluid expressed in terms of mass per unit volume. So the density of water, for example, is universally referred to as about 62 pounds per cubic foot in the English system. In the SI system, it is one kilogram per liter, which gives a clue how the size of the liter was originally picked.

Density can be expressed as pounds per cubic foot, as in the water example, but more commonly, especially for pipeline work, relative density, or *specific gravity*, is used:

Specific gravity = Density of fluid/Density of reference material

The density of pure methane is 0.554, meaning it has 55.4% of the mass of the same volume of air. Natural gas as transported on transmission and distribution pipelines in addition to methane contains a few other heavier molecules, giving it specific gravities in a range between 0.6 and 0.7. Specific gravity is often referred to simply as "gravity" by industry insiders.

Natural gas is sold at agreed-upon temperatures and pressures or standard temperature and pressure (STP). In the English system, STP is 60°F and atmospheric pressure about 14.7 psia; the SI system uses 0°C and atmospheric pressure of 100 kPa. Natural gas streams have similar, but not exactly the same, energy contents. Thus, each stream is also adjusted for energy content, measured in British thermal units (Btus) or therms. One Btu is the quantity of heat necessary to raise the temperature of 1 pound of water 1°F, from 58.5°F to 59.5°F, under standard pressure of 30 inches of mercury. One *therm* is equal to 100,000 Btus. One *dekatherm* is equivalent to 1,000 therms or 1,000,000 Btus.

The American Gas Association (AGA), the American Society for Testing and Materials (now ASTM International [ASTM]), and others have developed extensive tables for the conversion of natural gas to standard conditions in support of commercial transactions.

Viscosity. Physicists attest that molecules tend to attract each other. It is what keeps the world together. This attraction causes internal resistance to flow called *viscosity*. The higher the viscosity, the more energy it takes to move the fluid because of the internal resistance resulting from molecular action. Natural gas has a relatively low viscosity compared to some things moving on pipelines, like crude oil. Even though the viscosity is relatively low, however, it is an important hydraulic property when calculating pressure loss.

Viscosity is expressed in one of two ways by names that are easy to muddle up:

- ***Absolute viscosity.*** Sometimes confusingly called *dynamic viscosity*, absolute viscosity is a measure of the force needed to move a gas at a constant velocity through a tube or pipe.
- ***Kinematic viscosity.*** This is the ratio of absolute viscosity to fluid density. (*Kinematics* is the branch of mechanics that deals with motion.)

$$\text{Kinematic viscosity} = \text{Absolute velocity}/\text{Specific gravity}$$

Absolute viscosity is measured in centipoises, named after Jean Louis Poiseuille (1799–1869), who developed the concept. Centipoise is pronounced "centi-pwaz," whereas Poiseuille is essentially unpronounceable.

Kinematic viscosity is measured in centistokes, after the sensibly named Sir George Stokes (1819–1903), whose law of viscosity became the basis for modern hydrodynamics.

Absolute viscosity is measured in a laboratory by timing how long it takes for a given volume of fluid at a specified temperature to pass through the capillary tube in a viscometer.

As with density, viscosity is affected by temperature. Natural gas becomes more viscous as temperature goes up and less viscous as it goes down. Interestingly, hydrocarbon liquids like gasoline and diesel behave exactly the opposite, with viscosity decreasing as temperatures increase and increasing as they get colder.

Specific heat. Compression of natural gas, or any other gas, causes an increase in temperature. As the gas expands, it cools. Engineers calculate these changes using a property called *specific heat* or *specific heat capacity*, which is the amount of heat required to raise 1 pound of natural gas by 1°F. The SI value not surprisingly is the amount of heat required to raise 1 kilogram of natural gas by 1°C.

Actually two specific heats are employed: the specific heat at constant pressure and the specific heat at constant volume. The ratio of these two specific heats is used to calculate gas expansion and contraction.

Vapor pressure. The propane tank under an outdoor grill contains both vapor and liquid. As gas (vapor) is fed to the grill, more of the liquid vaporizes and maintains the pressure in the tank. When the last of the liquid evaporates, the pressure in the tank begins to drop. The amount of gas going to the burners declines to nothing as pressure inside the tank approaches atmospheric pressure.

The importance of vapor pressure in a natural gas pipeline has to do with the molecules other than methane. Ethane, butane, and propane are often present in small percentages. When the combination of natural gas pipeline operating temperatures and pressures exceeds the vapor pressure of these heavier molecules, they turn to liquid, potentially blocking flow and damaging compressors and other equipment.

Compressibility. All materials, solids, liquids, and gases, compress or expand when force is applied. Steel beams in skyscrapers actually compress a small amount because of the weight they carry, and engineers design accordingly. Steel pipelines stretch (stretch is called *strain* by engineers) and get slightly larger in diameter as pipeline pressures increase. They return to their original diameter when the line is depressurized, as long as they are not stretched too much, exceeding the *elastic limit* of the steel.

Compressibility is the measure of how much the volume of a substance (in this case gas) changes when a force is applied to it, compared to its original volume—in other words, how much work it takes to force a given mass into a smaller space. Compressibility is discussed in more detail in the next section.

Why gases behave that way

About the time Newton made his intellectual breakthroughs in motion and gravity, Robert Boyle (1627–1691) earned the title of Father of Modern Chemistry. In 1662, he discovered rules of gas behavior that became known as Boyle's law. In 1787, Jacques Charles (1746–1823) added to the knowledge about gases with his own gas laws. Joseph Louis Gay-Lussac (1778–1850) added a third law around 1801 to complete the picture. These three remarkable and universal relationships apply not just to some gases, but to all gases. There is, however, a caveat: the gas must be an *ideal gas*. The American Gas Association (AGA) defines an ideal gas as "the combination of the volume, temperature, and pressure relationships of Boyle's and Charles' laws resulting in the relationship $PV = RT$."[1] What this means in simple terms is that the concept of ideal gas neglects the intermolecular forces between the molecules.

Since no gas is ideal, a correction factor is necessary, called the *Z factor*. It accounts for the attraction and repulsion occurring between the gas molecules causing them not to always obey the ideal gas law. Since attraction and repulsion are partly dependent upon how close the atoms are packed together, the Z factor varies according to the pressure.

Where *P*, *V*, and *T* are the absolute pressure, volume, and temperature of a gas, respectively, Boyle's law says the following:

$$V_1/V_2 = P_2/P_1$$

That is, the change in the volume of a gas is inversely proportional to the change in its pressure, and vice versa. For the same amount of gas, raising the pressure decreases the volume as long as the temperature stays constant (fig. 3–7).

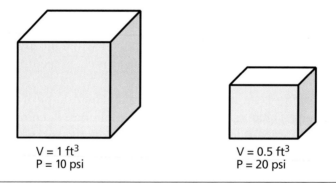

V = 1 ft³
P = 10 psi

V = 0.5 ft³
P = 20 psi

Fig. 3–7. Boyle's law. Reducing the volume by half doubles the pressure.

Then along came Charles's law, which states the following:

$$V_1/V_2 = T_1/T_2$$

That is, the change in volume of a gas is proportional to the change in its temperature. For the same amount of gas, raising the temperature raises the volume as long as the pressure remains constant (fig. 3–8).

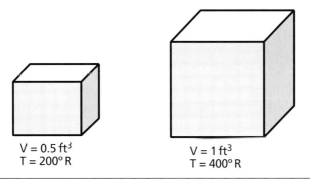

V = 0.5 ft³
T = 200° R

V = 1 ft³
T = 400° R

Fig. 3–8. Charles's law. Doubling the temperature doubles the volume.

Finally, Gay-Lussac's law states the following:

$$P_1/P_2 = T_1/T_2$$

That is, for a given volume of gas, the temperature will vary in proportion to the pressure. For the same amount of gas, raising the temperature raises the pressure as long as the volume remains constant (fig. 3–9).

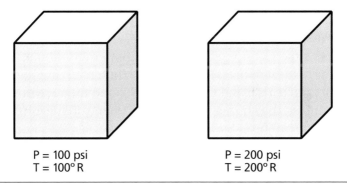

P = 100 psi
T = 100° R

P = 200 psi
T = 200° R

Fig. 3–9. Gay-Lussac's law. Doubling the pressure doubles the temperature.

These three laws have to do with how gases respond to the addition of temperature or pressure in ideal situations. Besides temperature, pressure, and volume, two more factors, the amount of gas measured in moles (expressed as n) and the universal gas constant (expressed as R) lead to the ideal gas law, which states the following:

$$PV = nRT$$

Since the ideal gas law only predicts gas behavior fairly accurately at low pressures, a compressibility factor, Z, as mentioned earlier, is used to compensate. Always less than 1, the Z factor "corrects" the fact that gas behavior deviates from ideal. Consequently, it is often called the *deviation factor*. Calculating the Z factor depends on the *critical temperature* (temperature above which the gas cannot be compressed into a liquid) and *critical pressure* (minimum pressure required to compress a gas at critical temperature into a liquid). Critical pressure and temperature and calculating the Z factor are beyond the scope of this book. Thankfully there are people (but not many) who actually enjoy talking about Z factors and handle the calculations for the rest of the world. The real gas law then is as follows:

$$PV = ZnRT$$

The important thing to remember about these laws is that pressures, temperatures, and volumes of any gas are intimately and rigorously interrelated. The real gas law and other hydraulic principles are used to develop *equations of state*, or equations aimed at modeling and predicting how the factors interrelate in specific applications. Equations of state are used in pipeline design and engineering, pipeline operations, and leak detection modeling.

Here is a practical operating point to remember about compressibility: as fluids are compressed, they absorb potential (pressure) energy, their temperatures rise, and they can lose heat to their environment. As they decompress, they release this potential energy in the form of kinetic energy, and absorb heat. Sudden pressure changes (like closing a valve or starting a compressor) generate pressure waves that travel along the pipeline, eventually dissipating like the rings of waves formed by a rock tossed into a pond.

The compressibility of gas makes it relatively forgiving of pressure surges. However, as in the balloon example, when gas lines leak, a tremendous amount of potential energy may be converted to kinetic energy very quickly, tearing the pipe, throwing soil and debris, and creating a distinctive roar.

These four hydraulic properties, density, viscosity, vapor pressure, and compressibility, are fundamental to the design and operation of natural gas pipelines. Equations of state, mentioned earlier, use these properties to predict fluid behavior. Literally thousands of charts and graphs have been developed through calculations and experimentation to represent the behavior of fluids over the range of conditions pipelines encounter. The next section discusses how these properties are used to analyze, predict, and control pipeline operations.

Hydraulics

Hydraulics is the engineering approach to fluid mechanics. In the 1860s, practical entrepreneurs built small and short pipelines. The hydraulic principles had been discovered well before the beginning of the gas and oil pipeline era, but early pipeliners were generally technologically challenged. The materials and equipment needed for long-distance, high-pressure pipelines were just being developed, and in the early days, much of the engineering acumen came from trial and error rather than formal technical training. During the early part of the 20th century, practical experimentation merged with theory and empirical research to produce the discipline of pipeline engineering.

Basic flow principles and equations

Most of what needs to be known about hydraulics comes from the laws of thermodynamics and Bernoulli's principle.

The first law of thermodynamics. This principle, also commonly called the *law of conservation of mass and energy*, states that matter or its energy equivalent cannot be created or destroyed. It can only remain as it is or change from one form to another. This notion had been around for some time when, in mid-1800s, James Prescott Joule (1818–1889) demonstrated the law through a series of experiments. (In his honor, the unit of energy in the metric system is called the *joule*.)

The first law is fundamental to pipeline hydraulics. The force of compressors puts energy into the gas, pushing it through the pipeline. Friction converts part of that energy to heat energy. Keeping track of these energy flows allows pipeline engineers to design and operate pipelines.

Bernoulli's principle. Another critical law regarding the behavior of fluids is Bernoulli's principle. In the 18th century, while trying to determine how to measure blood pressure, Daniel Bernoulli (1700–1782) inserted a straw into a pipe and noted that the height to which the liquid in the straw rose was directly related to the speed of the fluid in the pipe. With faster flow the liquid level in the straw dropped; with slower flow it rose. The principle has broad application, especially to pipelines. After more experimentation and analysis, Bernoulli mathematically described the relationships thereafter known as the Bernoulli principle. In its purest form, it states the following:

$$\text{Static pressure} + \text{Dynamic pressure} = \text{Constant}$$

Static pressure is the potential energy (pressure) contained in the pipeline. *Dynamic pressure* is the kinetic energy of the fluid, the energy it has because it is moving. The amount of kinetic energy depends on the velocity and density of the fluid.

Figure 3-10 demonstrates Bernoulli's principle. The lower velocity of flow in sections A and C result in lower dynamic pressure than in section B, so the static pressure (as measured by the liquid in the vertical section of pipe) is higher in sections A and C than section B. Note the velocities are different, but the flow rates in mass per unit time are the same in each section.

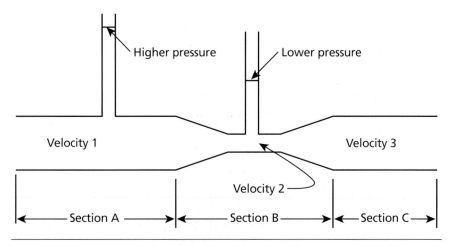

Fig. 3–10. Bernoulli's principle. Velocity 1 = Velocity 3, Velocity 2 is greater than Velocity 1, and the pressures in sections A and C are higher than in B.

Based on this principle, Bernoulli developed several equations to explain the behavior and flow of ideal fluids in ideal situations. These equations became the basis of more practical equations developed over time and through experience. Curiously, as experience accumulated over the last 100 years, engineers found some equations are more convenient for predicting results for larger diameter pipe and others for smaller diameter pipe, and some are unique to the fluids. Not much space in this book is devoted to flow equations, but it is important to know a bit about the following:

- Flow characteristics
- Friction loss
- Elevation loss (or gain)
- Flow rates

Flow characteristics

Fluid flow through pipelines is either laminar, transitional turbulent, or fully turbulent. Knowing which of these is happening is important because it affects pressure loss calculations.

Laminar flow is pleasantly smooth, streamlined flow. Each molecule marches along in a (relatively) straight line. It is pushed by the molecule behind and pushes the molecule ahead, seldom bothering its neighbor to the left, right, top, or bottom. It is called *laminar* because the molecules move in layers over each other. A flag gently waving in the breeze is an example of the air (a mixture of gases) in laminar flow as each molecule moves peacefully past the flag (fig. 3–11).

Fig. 3–11. Flag with air in laminar flow. The flag hangs limply as the air moves straight by in laminar flow.

In laminar flow the layers move at different speeds, with those near the center of the pipe moving the fastest and those at the wall theoretically not moving at all (fig. 3–12).

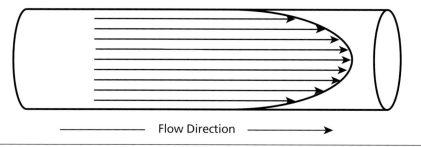

Fig. 3–12. Laminar flow profile

Continuing the flag analogy, in *turbulent flow* the air molecules bounce around in multiple directions, striking the surface of the flag, bouncing off and forcing the flag over such that the flag strikes other molecules, pushing them around. Eddy currents set up along the flag, causing it to snap and flutter, rather than wave, as air molecules careen along the flag's surface (fig. 3–13).

Fig. 3–13. Flag with air in turbulent flow. The flag snaps briskly as the air molecules strike the flag and bounce off, setting up eddy currents.

Turbulent flow through gas pipelines happens the same way. The inside of plastic or steel pipe may look smooth, but to molecules even a little unevenness makes a big difference. If shoved against the side by their neighbor, they bounce off the wall, colliding with the same or other neighbors, jostling all around (fig. 3–14).

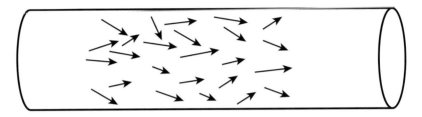

Fig. 3–14. Turbulent flow profile

Most high-pressure transmission pipelines operate in turbulent flow, but distribution lines at any time can be in laminar, transitional turbulent, or fully turbulent flow, or not flowing at all. Which flow regimen the line is operating in is unimportant to the consumer, but engineers must understand the flow regimen and calculate friction loss and flow rates accordingly.

Sir Osborne Reynolds (1842–1912) is widely credited with discovering laminar flow around 1880. He injected dye into a stream of fluid and found that at lower flow rates, the flow remained streamlined. At higher velocities, the flow went from streamlined to turbulent. In 1882, he published a paper regarding this topic and introduced a number called, not surprisingly, the *Reynolds number*. Reynolds numbers are calculated based on density, viscosity, velocity, and diameter. The size of the Reynolds number predicts whether flow is laminar or fully turbulent. Reynolds numbers appear prominently in pressure loss equations.

Friction loss

For most pipelines, friction loss and not elevation change is the major factor requiring compressor stations. Friction loss is caused by the molecules rubbing against each other and against the pipe wall. It is affected by the following:

- Viscosity
- Density
- Velocity (a function of pipe diameter and flow rate)

- Pipe length
- Roughness of the inside of the pipe
- Changes of direction caused by turns in the pipe

The unpredictable nature of molecules bouncing around in turbulent flow makes modeling the behavior of individual molecules impossible. But over the years, hydraulic engineers, based on experimentation tempered with practical knowledge, have devised shortcuts to calculate friction and pressure loss. (As in the garden hose examples, *friction loss* is energy loss, and that translates to pressure loss.) Charts and graphs based on fine-tuned experimentation are fed into simplified equations to give practical and acceptable results. A variety of hydraulic equations, normally named after the person or organization developing them, are currently in use. Each has its own application based on diameter, pressure, range of operating variables, and other more obscure factors such as how close the fluid is to its dew point. Pressure loss and flow calculations became more efficient as computer models were constructed.

As natural gas moves along the line, pressure is reduced by friction. The natural gas farther downstream of the compressor is at a lower pressure than near the compressor (assuming the same elevation). Since natural gas is compressible, this loss of pressure allows the natural gas to uncompress (expand), decreasing density, increasing velocity, and therefore increasing friction loss. Natural gas pipeline modelers take care to include this phenomenon as they calculate pressure loss due to friction.

Elevation loss (or gain)

As mentioned earlier, pressure decreases as fluids move uphill and increases as they flow downhill, but how much? That depends on the weight of the fluid and the height of the hill. The heavier the fluid, the more pressure it takes to push it up the hill. Higher hills require more pressure than lower ones.

To use another water example, a cube of water with dimensions of 1 foot by 1 foot by 1 foot (1 cubic foot) weighs 62.4 pounds (fig. 3–15).

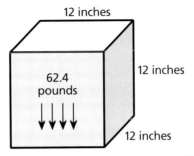

Fig. 3–15. One cubic foot of water. The pressure exerted is 0.433 psi.

The entire 62.4 pounds, as a column, is pushing down on an area 12 inches × 12 inches square, or 144 square inches.

The pressure at the bottom of the cube is given as follows:

62.4 pounds/144 square inches = 0.433 pounds per square inch

If the 12-inch square column of water is 100 feet tall, the pressure at the bottom is given as follows:

0.433 psi/ft × 100 ft = 43.3 psi

The pressure at the bottom of the column is 43.3 psi greater than the pressure at the top of the column.

Gas works the same way, but with an important difference. Remember, liquids are essentially not compressible, so their density does not depend very much on pressure. Each successive cubic foot of water up the column weighs the same as the one under it. However, this is not true for gas. Since gas is compressible, each successive cubic foot of gas up the column is less dense than the one below. It weighs less than the one underneath.

Pressure caused by gravity depends on the height of the column of fluid, not the geometry of the column (fig. 3–16). It may seem the pressure should be higher at the bottom of the slanted pipe on the left because there is a longer column, but it is the same at the bottom of both pipes. The *total* force exerted is greater for the pipe on the right because the pressure is exerted over a greater area at the bottom, but not the force per square inch, or the pressure.

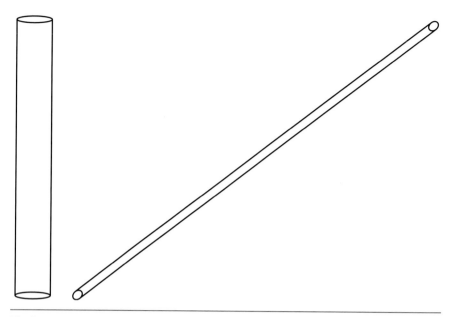

Fig. 3–16. Pressure at the bottom of two pipes. Pressure per square inch due to gravity at the bottom of both pipes is the same. Total force at the bottom of the larger pipe is greater because of its greater surface area.

Flow rates and capacities

The basic flow equation says flow is equal to the density multiplied by the velocity of movement past a point, multiplied by the cross-sectional area through which the material is moving, multiplied by the amount of time it is moving past the point (fig. 3–17). Accordingly, flow is usually calculated from velocity, density, and time, rather than measured directly.

Flow rates come in two varieties: the current flow rate and the ultimate capacity. Natural gas transmission lines servicing cities must have sufficient capacity to serve the town on the coldest day of winter and the local gas-fired power plants on the hottest day of summer. They normally flow well below their ultimate capacity in the off-season (fig. 3–18). Local distribution lines normally flow faster in the morning, as residents awaken, shower, and cook breakfast, than at night, when most people are asleep.

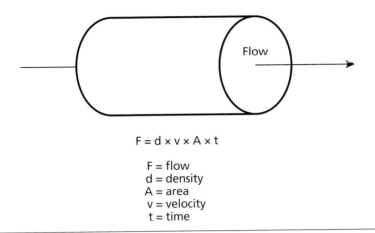

Fig. 3–17. Basic flow equation. Flow equals the amount going past a point over time. The amount is equal to density times velocity times the cross-sectional area times the amount of time.

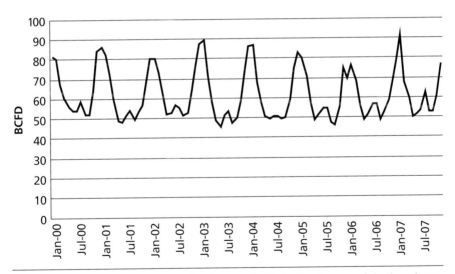

Fig. 3–18. US natural gas consumption by month. Note the sharp peak in the winter driven by the need for heat and the small peak in the summer driven by electrical power generation.
Source: Prepared by Miesner LLC from US Energy Information Administration data.

The input into gas lines is more uniform than the output. Gas wells produce into gas lines at a relatively constant rate over days and years. At the other end, consumers generally use more gas during the day than at night, and more in the winter than in the summer. The amount of gas in the line is known as *line pack*. Since gas is compressible, on a short-term basis more can be put into the line than is removed, a process known as *packing the line*. The converse is also true and not surprisingly is known as *unpacking the line*. Gas lines can go from trough to peak demand over a short period of time. Gas storage and line pack are used to help level off this imbalance. Week-to-week and season-to-season demands can vary 50%, and storage and compression capacity usually are used to handle these variations (fig. 3–19).

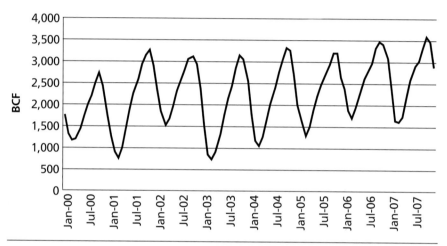

Fig. 3–19. US working natural gas in storage. Storage fills up in the summer when demand is lowest and is drawn down in the winter months.
Source: Prepared by Miesner LLC from US Energy Information Administration data.

Summary

- Potential energy (pressure) is the motive force making gas flow.
- The basic flow equation says flow is equal to density times velocity times cross-sectional area times time.
- Friction loss uses up pressure. (Actually, it converts it to other forms of energy like heat.)
- Friction loss depends on fluid properties, velocity, and pipeline configuration, including diameter, length, and turns.
- Equations of state describe fluid flow.

- Multiple equations of state exist, all based on the same fundamental laws of physics.
- Equations of state are used to design, operate, and control pipelines.

Note

1. Extracted March 24, 2016 from https://www.aga.org/knowledgecenter/natural-gas-101/natural-gas-glossary/i.

chapter 4

Components and Equipment

Plastics.

—*Mr. McGuire in* The Graduate

Introduction

When most people think of natural gas distribution systems, they think of what they see: the meter and pressure regulator connected to their house. They do not spend much time thinking about all the other pieces of equipment and components comprising the system. Nevertheless, items such as pipes, fittings, flanges, bolts, nuts, and so forth are also clearly components. The word *equipment*, as used in this text, is an assemblage of components. *Skids* are normally steel I-beams welded together, often with a steel decking. *Provers* are used to calibrate meters. Meters, odorant injection skids, and provers are normally thought of as pieces of equipment.

There are also all the in-between items, such as valves, pressure control systems, and metering skids. Meters, valves, instruments, fittings, and pipe are often assembled and mounted on a skid in the factory and then hauled to the site. Skid mounting allows easy installation and facilitates moving the entire assemblage to another location. Whether equipment or component, all the items must work together in harmony to safely and efficiently deliver gas to every home, office, hospital, school, and place of business at exactly the proper pressure every day, all day.

Pipe

Early distribution lines were made from a variety of materials, including lead, copper, cast iron, and even musket barrels connected together. By the dawn of the 20th century, cast iron had become the piping material of choice for distribution lines and service connections. Next came steel, and then around the late 1950s, plastic pipe entered the arena.

According to the American Gas Association, 55.3% of the distribution mains in service in 2012 in the United States were plastic pipe, 44% were steel, less than 3% were cast or wrought iron, and there was a small percentage of "other." Service connections, also according to AGA for the same year, were even more heavily weighted toward plastic. Nearly 68% were plastic, about 40% were steel, about 1% were copper, and 2% were "other."[1]

The split between plastic, steel, and even cast iron varies regionally. Cities that built a large percentage of the distribution infrastructure prior to about 1960 tend to have greater concentrations of steel and cast iron mains and services than cities that expanded after that time.

Finding statistics for other countries is more difficult, but it is reasonable to believe the percentage for western Europe is similar to the United States, owing to their similar development profile. The natural gas distribution systems outside the United States and western Europe developed more recently, so the mains and service lines are more likely to be plastic pipe, which first came into use in the mid- to late 1950s. Nearly all new mains and service connections operating at less than 100 psi are plastic, either high-density or medium-density polyethylene (HDPE or MDPE). HDPE is more prevalent, owing to its higher pressure carrying capacity.

Natural gas transmission lines, which deliver natural gas to the distribution companies, are primarily constructed from steel pipe. Steel has greater strength than plastic for the same wall thickness, making steel an ideal choice for long-distance gas movements and the higher operating pressures those longer movements require. As higher strength plastics and various composite pipes develop, the percentage of transmission lines comprised of steel likely will decline slightly, although not rapidly.

Plastic pipe

Plastic pipe displaced steel pipe as the material of choice for the distribution industry since it does not corrode and is cheaper to manufacture and install than steel. At smaller diameters, plastic pipe can also be *squeezed off*, whereby pipe walls are squeezed together with a crimping tool to stop flow without using a valve. This is a boon for maintenance and repairs.

To make the pipe, polyethylene pellets are melted, and the resulting hot liquid is extruded through different size dies, depending on the desired diameter. While it cools, the extruded plastic pipe is pulled through sizing plates, ensuring it retains the specified diameter. The cooled and solid pipe is then cut to length and labeled at frequent intervals. The labeling usually includes the nominal pipe size, type of plastic, and the *standard dimension ratio* or SDR, which is the ratio of the outside diameter to the wall thickness. It can also be labeled with the pressure rating, the manufacturer's name or trademark, and a manufacturing code (fig. 4–1).

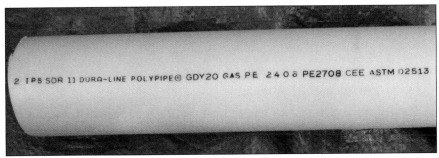

Fig. 4–1. Two-inch nominal pipe with an SDR code of 11, meaning the outside diameter is 2.375 inches, and the wall thickness is at least 0.216 inches.

Pipe properties. ASTM International (formerly the American Society for Testing and Materials) publishes standards for pipe properties. The latest standard for PE pipe is Standard D 3350-14, "Standard Specification for Polyethylene Plastics Pipe and Fittings Materials." According to the standard, "Classification of materials shall be according to tests of primary properties like density, melt index, flexural modulus, tensile strength at yield, slow crack growth resistance, and hydrostatic strength classification."[2] Each of these properties is considered when selecting the proper pipe for the specific application.

Operating pressures. As pressure builds in a pipeline, the pipe gets a little bigger around. Engineers call this stretching *strain*, a measure of change in length, in this case diameter. The force (pressure) making the pipe stretch is called *stress*. Pipe, just like a rubber band, is elastic within certain stress ranges. As long as stress remains below the specified minimum yield stress (SMYS) of the material, in this case plastic, the pipe, like the rubber band, returns to its original size when the pressure is relieved. If stretched too far and subjected to too much stress, however, the pipe gets permanently bigger around, meaning the pipe wall gets thinner. If the pipe wall becomes too thin, its strength is reduced; stretched too far, it will burst.

Engineers consider the pipe's properties and its wall thickness and diameter and then include a safety factor when calculating safe working pressures. Care must also be taken to understand how much force the material over the pipe will exert on the buried pipe and how much the pipe will deflect. This is particularly true at road crossings and other crossings.

Color coding. Plastic pipe is used for many purposes, so it is color coded according to the industry, using the following convention:

- *Yellow.* Gas, oil, steam, petroleum, or gaseous materials.
- *Red.* Electric power lines, cables, conduit, and lighting cables.
- *Orange.* Communication, alarm, or signal lines, cables, or conduit.
- *Blue.* Water, irrigation, and slurry lines.
- *Green.* Sewers and drain lines.
- *Purple.* Reclaimed water lines.

Accordingly, plastic pipe used for natural gas is most commonly yellow, although black pipe with yellow stripes is also allowed. A word of caution, however, is necessary: some plastic pipe went into service before color coding became common, so all pipe must be treated with care.

Smaller plastic pipe, usually 6 inches or less in diameter, is often produced in lengths of several hundred feet and wound onto a spool. The pipe can be unwound later at the work site (fig. 4–2).

Fig. 4–2. Spools of HDPE pipe stacked in a storage yard

Spooling several hundred feet of pipe limits the number of connecting welds required, speeding installation time and reducing the potential for weld defects.

Larger diameter plastic pipe is produced in lengths that are welded or fused together end-to-end during the construction process (fig. 4–3).

Fig. 4–3. Lengths of HDPE pipe at a job site

Final thought. As the use of plastic pipe is relatively more recent than the use of steel, there is a question as to the useful life of plastic pipe, an issue receiving considerable study by industry groups. One excellent resource to learn more about plastic pipe is the Plastics Pipe Institute (http://plasticpipe.org).

Steel pipe

Steel pipe used to construct cross-country pipelines, commonly referred to as *line pipe*, is produced in standard sizes and strength ratings. Incongruously, the outside diameters of 4-, 6-, 8-, 10-, and 12-inch line pipe are 4.5, 6.625, 8.625, 10.750, and 12.750, respectively. Starting at 14 inches, the outside diameters are whole numbers: 14, 16, 18, and so on, in 2-inch increments. In any event, standards agreed to by line pipe manufacturers and pipeline companies list the possible diameters and corresponding wall thicknesses. Note that line pipe is not limited to cross-country pipelines. When line pipe is used to construct stations, it is simply cut to whatever length is needed.

Manufacturing methods. Most steel pipe is made by one of two methods, either piercing or bending. In the 150-year-old piercing process, invented by the Mannesmann brothers, a super-hot steel cylinder is pushed over a *mandrel* (a solid rod) that hollows out the cylinder, forming a seamless pipe. The cylinder is further rolled to ensure roundness, the ends are trimmed for a square fit, and then the insides are smoothed.

The most common production method for steel pipe used in natural gas service, however, is bending a flat steel plate, commonly called *skelp*, into a cylinder and welding the seam longitudinally to form *welded pipe*. Over the years, the longitudinal weld seam has been made by several methods:

- Furnace butt welded
- Furnace lap welded
- Electric flash welded
- Electric resistance welded (ERW)
- Double submerged arc welded (DSAW)

The first three weld seam methods have been displaced over time by the last two, ERW and DSAW. DSAW pipe is also called SAW which is shorter than DSAW. In either case, both DSAW and SAW have two longitudinal weld seams, one on the inside and the other is on the outside.

ERW pipe manufacturing involves forcing the two edges together and inducing current to the pipe. The current heats the two edges, melting the steel to bond them together. No "filler material" is used for ERW pipe; the two sides are just pressed together. After the pipe cools, the weld is trimmed, so it is often hard to see the actual weld seam.

SAW (or DSAW) adds filler material from a coil of wire placed above the edges of the skelp to serve as an electrode. An electrical source is connected to the filler material and the pipe. Current jumps across the gap from the filler material to the skelp, forming an arc. The arc heats the edges of the skelp and at the same time melts the filler material. The filler material slowly unwinds and melts as the skelp is pushed past the spool, resulting in a smooth and uniform weld seam, slightly protruding above the pipe's outside diameter (fig. 4–4).

The smallest diameter in which DSAW pipe is produced is 14 inches, so DSAW pipe sees limited use in natural gas distribution systems but is widely used in transmission systems.

Fig. 4–4. DSAW weld seam

The need to move more natural gas led to a twist on forming pipe from flat skelp to make a *spiral wound pipe*. Rather than folding the plate into a tube, the flat steel is twisted like a toilet paper roll. The edges are then welded together (fig. 4–5).

Fig. 4–5. Spiral wound pipe

Recalling from high school geometry the circumference of a circle is pi times diameter leads to realizing twisting the flat steel into a spiral means the skelp no longer needs to be as wide to produce the same diameter pipe.

In each case, after the pipe is welded, it is cut to specified lengths. Each manufacturing method also includes postwelding treatment to relieve stress and to size and straighten the pipe. During the manufacturing process, the pipe undergoes numerous checks to ensure quality.

Pipe properties. The American Petroleum Institute publishes API 5L, *Specification for Line Pipe*, one of the oldest and most frequently used standards in the pipeline industry. It specifies chemical and physical properties for the parent metal and the pipe, and also the testing to verify those properties.

Other than iron, carbon is one of the key components of steel pipe, affecting its strength, ductility, and other properties. Other chemicals, including manganese, phosphorus, sulfur, columbium, chromium, copper, molybdenum, vanadium, silicon, and titanium are also added as needed to achieve special properties. Pipeliners are interested in the chemical properties of the metal mainly as they affect the performance of the pipe. Strength, ductility, fracture toughness, and ease of welding are important considerations and are tested according to industry standards.

Specified minimum yield strength (SMYS), the lowest point at which the pipe begins to yield, is the pipe property that gets the most attention. Pipe is produced in various levels of SMYS; for example, X-42 has an SMYS of 42,000 psi, X-46 has an SMYS of 46,000 psi, and so forth, all the way up to X-80 or even higher. SMYS, diameter, wall thickness, and a safety factor all go into calculating safe operating pressures.

Steel pipe is commonly sold by weight; half-inch wall X-42 costs roughly the same as half-inch wall X-80, but X-80 can safely hold substantially higher pressures. So why would someone not just buy the strongest, thinnest wall pipe possible? The metallurgy required to achieve the higher strength pipe also requires more care in welding than lower strength pipe, and it can also be less ductile (more brittle). Thinner wall pipe may also have a tendency to "oval" if the ratio of wall thickness to diameter (D/t) gets too large.

Coatings

Unprotected steel rusts and so do pipelines, which is a disadvantage of steel pipe compared to plastic pipe. Corrosion is caused by current flow, so steel pipelines are externally coated to insulate the line electrically.

The most common types of pipeline coatings for corrosion control include fusion bond epoxy, coal tar enamel, various forms of extruded plastics, and various types of cold wrap tapes. In the 1980s shrink sleeves became popular but have since been discontinued owing to installations problems. Many steel pipelines constructed during that time still have shrink sleeves however. Fusion bond epoxy has largely replaced the other coatings for new construction because of its superior properties. The girth welds are then coated, normally with cold applied epoxy to complete the continuous coating.

Fusion bond epoxy

Fusion bond epoxy (FBE) is a powder that is sprayed onto the pipe surface after the pipe has been cleaned and heated to more than 450°F. The epoxy powder melts onto the steel surface and fuses to the pipe, creating a tough, insulating barrier. Most FBE applications take place at a coating facility (fig. 4–6).

Fig. 4–6. Pieces of 4-inch FBE-coated pipe

Coating factories stop the coating about an inch or so from the end of each pipe to facilitate *circumferential* welding. After the pipe is welded, the weld must be coated. One girth weld coating method is painting on a two-part, cold-applied coating (fig. 4–7).

Fig. 4–7. Coated girth weld

Coal tar enamel

Coal tar enamel (CTE) coating has been used in various forms for nearly 100 years (fig. 4–8).

Fig. 4–8. Early pipe coating process
Courtesy: Association of Oil Pipelines.

CTE is a thick residue that comes as a by-product from coal coking ovens. First a primer is applied, then coal tar, then inert fiber fillers are added to improve performance properties, and finally fiberglass mats are embedded as a wrap to give added strength and stability. Any skips in the primer or voids under the coating create areas of potential corrosion and must be strictly avoided. Owing to its long use, more miles of pipeline worldwide are coated with CTE of one form or another than any other coating. As a note of caution, some CTE coatings contain asbestos, so appropriate precautions are needed when working around CTE coatings.

Tapes

Before the advent of FBE, tapes were used to coat girth welds on pipes coated with coal tar enamel. Tapes are also sometimes used to coat girth welds joining FBE-coated pipe. The tape is supplied in rolls of varying widths, depending on the diameter of the particular piece of pipe, with wider tapes for larger diameters. They are spiral wrapped on the pipe, overlapping the plant-applied coating. Each subsequent wrap of the spiral overlaps the previous wrap to form a tight seal (fig. 4–9).

Tapes are also used extensively to repair damaged coating. The damaged or loose coating is removed, and the pipe is cleaned off. Tape is spiral wrapped to cover the damaged area, taking care to overlap sound coating to ensure a good seal. Thorough cleaning and careful application is always necessary to ensure no voids or holes remain that could concentrate current flow and create corrosion hot spots.

Fig. 4–9. Tape wrapped over girth weld on FBE-coated pipe

As a note of caution, sometimes tapes or other coating materials meant to shield the pipe from external currents *disbond* (come loose from the pipe). In these circumstances, not only will these coatings or tapes no longer prevent corrosion, they may now prevent the rectifier from forcing current flow in the desired direction. The result may be a corrosion hot spot.

Shrink sleeves

Shrink sleeves are polymers that come in precut sections to fit around a particular diameter pipe. The pipe is cleaned off and the shrink sleeve is applied. As it is carefully heated, the sleeve shrinks to conform tightly to the joint profile. In the 1980s and 1990s, shrink sleeves were frequently used to cover weld areas during new construction or to replace damaged coating.

Over time, significant disadvantages to the use of shrink sleeves became apparent. Great care is required to apply shrink sleeves, and uniform heating is difficult to achieve, particularly on the bottom of the pipe. Consequently, shrink sleeves were sometimes not installed correctly, creating voids where corrosion could form. Most companies no longer allow shrink sleeves, but there are legacy installations on many pipelines constructed in the last part of the 20th century.

Abrasion resistant overcoating

Pulling coated pipe through directional drills can damage the coating. First concrete coating was used to protect the primary coating, and then abrasion resistant overcoating (ARO) was developed. ARO is a hard, mechanically strong top coating applied over fusion-bonded epoxy pipeline corrosion protection coatings. It forms a tough outer layer that resists gouges, impact, abrasion, and penetration, protecting the primary corrosion coating from damage during pipeline directional drilling applications, bores, river crossings, and installations in rough terrain. This type of coating system is also known as *FBE dual coating*.

Concrete coating

The advent of ARO meant the use of concrete-coated pipe dropped significantly. Concrete coating is still used to add negative buoyancy (weight) to pipelines in high water table terrains, such as swamps, flood plains, and river crossings, to keep the pipeline from floating out. It is installed over the primary corrosion coating and formulated to allow some flexibility in order to avoid cracking during handling and installation (fig. 4–10).

Fig. 4–10. Concrete-coated pipes

Concrete coating is applied over weld joints at the work site after the girth weld is completed and the cathodic coating is installed.

Fittings

Pipelines constantly change direction and sometimes diameter, and they may have hundreds, even thousands, of connection points. In many cases, the plastic pipe itself is flexible enough to accommodate direction changes. Steel pipe is considerably less flexible, but it can be bent on-site with a pipe bending machine or at a factory with specialized equipment. When natural flexibility and bending is not sufficient, factory-produced fittings such as 90s, sweep elbows, 45s, reducers, and other shapes are installed to facilitate changes in direction or diameters. Tees and wyes connect one pipe to another, such as a service line to a distribution main, and reducing tees or wyes connect smaller lines to larger ones. Fittings are classified according to the following:

- *Composition material.* Plastic, steel, or cast iron.
- *Function.* Elbow (90, 45, sweep), tee, reducer, or coupling.

- *Method of connection.* Screwed, welded, fused, compression, bolted, or clamped.

Plastic fittings

Plastic fittings are made by injecting melted plastic into molds or by joining sections of pipe, machined blocks, or molded fittings together to produce the desired configuration. The ASTM list of six plastic pipe properties previously discussed applies equally to fittings. Safe operating pressures are determined by these properties, along with wall thicknesses and diameter.

Injection molded fittings. Plastic pellets are melted, and the resultant fluid is forced into a mold. When the fluid solidifies, the mold is removed, leaving the now-solid plastic in the shape of the mold. Typical injection molded fittings include tees, 45° and 90° elbows, reducers, couplings, caps, flange adapters and stub ends, branch and service saddles, and self-tapping saddle tees (fig. 4–11).

Fig. 4–11. Injection-molded reducers and 90° elbows

Thermoformed fittings. A section of plastic pipe is immersed in a hot liquid to make it pliable. When the pipe is at the proper temperature, it is removed from the hot liquid and shaped with a tool. The shape is then held until the pipe is cool. Some examples are elbows, swage reducers, and forged stub ends. Sometimes thermoformed fittings join together injection-molded fittings or parts with sections of plastic pipe to produce a desired shape.

Either *butt welding*, heating two ends and pushing them together, or *socket welding*, heating two pieces and pushing one into the other, are typical joining techniques for both injection-molded and thermoformed fittings.

Electrofusion fittings. Electrofusion couplings and fittings add one more step to the manufacturing process. Coil-like integral heating elements are installed into the injection-molded or thermoformed fittings. During installation, an electrofusion machine is connected to the two ends of the heating element. Electricity passing through the coil heats the coil, melting some of the plastic, which then fuses with the plastic on the mating part. This process takes the place of butt welding or socket welding (fig. 4–12).

Fig. 4–12. Electrofusion tee. Note the two connection points, one on each side of the tee.

Specifications for socket, butt fusion, and electrofusion fittings have been developed by ASTM and include the following:

- D 2683, "Standard Specification for Socket-Type Polyethylene Fittings for Outside Diameter-Controlled Polyethylene Pipe and Tubing"
- D 3261, "Standard Specification for Butt Heat Fusion Polyethylene (PE) Plastic Fittings for Polyethylene (PE) Plastic Pipe and Tubing"
- F 1055, "Standard Specification for Electrofusion Type Polyethylene Fittings for Outside Diameter Controlled Polyethylene and Crosslinked Polyethylene (PEX) Pipe and Tubing"
- D 2657, "Standard Practice for Heat Fusion Joining of Polyolefin Pipe and Fittings"

Compression fittings. Used to connect two plastic pipes together, *compression couplings* have three parts: two external collars and an internal piece. First one external collar slides over each pipe, and one end of the internal piece is inserted into each pipe. Then the collars are squeezed toward the center, compressing the internal piece against the inside of the pipe and the collar against the outside of the pipe (fig. 4–13).

Fig. 4–13. Two couplings are used to insert a short segment of plastic pipe, replacing a piece damaged during fence construction.

Steel fittings

Some steel fittings are welded to the pipe, others are screwed to it, others are clamped to it, and still others use the force of friction imparted through compression to hold on to it. Like plastic fittings, steel fittings come in a variety of shapes and have their own pressure ratings.

Screwed fittings. The fittings and short pieces of pipe (nipples) have threads, either internal or external, cut into them, and the various pieces are screwed together. House meter assemblies, for example, normally consist of the meter, a pressure regulator, and valves, all connected together with screwed fittings (fig. 4–14).

Fig. 4–14. Screwed fittings used in a standard house connection

Screwed fittings are often used for connecting lower pressure, smaller diameter pipes and components.

Flanges. Sometimes lumped together with fittings and sometimes discussed separately, flanges frequently connect components into steel piping systems. Flanges come in various configurations: weld neck, slip on, raised face, flat face, ring joint, and blind, just to mention a few. Each has a specific application. One of the more common flanges is the raised face weld neck flange (fig. 4–15).

Flanges frequently connect higher pressure, larger diameter pipe and components to each other.

The flange in the center of figure 4–15 has a raised face. Placing a gasket between the raised face on the flange and the mating component ensures a snug fit and no leaks (fig. 4–16).

Fig. 4–15. Collection of fittings and flanges, some of which have already been welded together

Fig. 4–16. Two-inch, 300-pound ANSI gasket

Welded fittings. Welding is another common connection method used extensively for higher pressure and larger diameter steel pipe and fittings. Figure 4–15 showed an assortment of welded fittings. The ends are usually beveled in a **V** shape, with the top of the **V** toward the outside so a welding rod can reach the inner surface, allowing a weld bead at least as thick as the wall.

Other fittings

Most items have a miscellaneous category, and fittings are no exception. Some are plastic, some are steel, and some are a composite of both plastic and steel. One common example is the transition piece, which is manufactured to facilitate field connections of plastic and steel pipe. It simply consists of a short piece of plastic pipe already connected to a short piece of steel pipe. Transition risers, for example, facilitate meter installations (fig. 4–17).

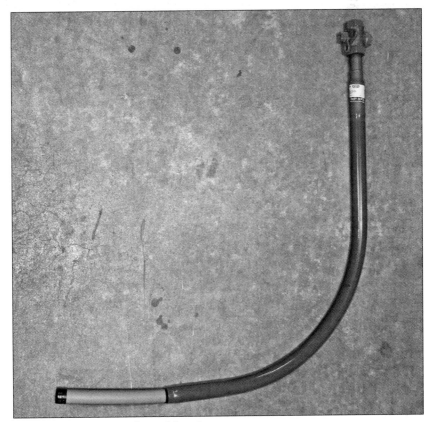

Fig. 4–17. Plastic-to-steel transition riser

The short section of plastic pipe on the fitting is either fused to the plastic service line or connected to it with a compression fitting. The short section of screwed steel pipe is connected to the steel end and serves as the start of the meter installation.

Small leaks are sometimes temporarily repaired by clamping a split steel sleeve lined with elastomer materials around the leak location. The bolts are then tightened, thereby pressing the elastomer material against the leak (fig. 4–18).

Fig. 4–18. The leak clamp is at the bottom of the figure on the vertical pipe. Note the soap bubbles on the top of the clamp, indicating the leak was not successfully stopped. In this case, the entire riser was replaced with a transition riser.

There are many other fittings not discussed here, and each is designed for a specific purpose.

Valves

As with faucets in home plumbing systems, valves open to allow flow to begin or shut to stop flow. As a faucet opens, the amount of flow increases because the larger opening requires less pressure (energy) to push fluid through, and thus more flows through. The reverse occurs as a faucet is closed. The amount of pressure required to push the fluid through increases, and consequently, flow decreases. *Throttling* (creating pressure restrictions) is the primary flow mechanism in natural gas distribution lines.

Valves come in several types, each of which performs one or more functions. The most common valve types used in natural gas distribution are the following:

- Plug
- Ball
- Gate (slab and wedge)
- Globe (in line and angle)
- Check (swing, ball, and diaphragm)

Plug valves

The main internal component of a plug valve is a tapered plug with a hole through it. Turning the plug so the hole is in line with the pipeline allows flow. Turning it perpendicular to the pipeline stops flow (fig. 4–19).

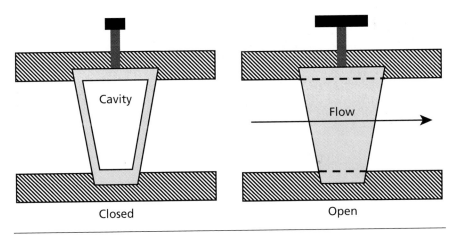

Fig. 4–19. Plug valve diagram

The primary function of plug valves is isolation—allowing flow or shutting it off. They are normally either open or closed. The tapered hole in the plug, wider at the top and narrow at the bottom, means the opening in the plug is smaller than the interior diameter of the pipe, so plug valves are not used in lines that require transit of a cleaning or inspection device. Plug valves are the most prevalent isolation valve type in service connections and many distribution mains.

Some larger diameter plug valves use the tapered feature of the plug to add an additional sealing element. An actuator (handle) is designed to rotate the valve to a closed position and then push down on the plug, forcing it tightly against two seats, one on each side, something like forcing a doorstop under a door. To open the valve, the plug is first raised to pull it away from the seat and then rotated 90°.

Since the plug is tapered, it can be double block and bleed. Forcing the tapered plug down against the seats forms two tight seals, hence the name *double block*. For the bleed feature, since both sides of the plug have tight seals, the inside cavity (when the hole is perpendicular to the flow) can be opened through a tap. Why is that useful? Emptying the cavity and verifying that flow stops once the cavity is empty indicates no fluid is leaking across the valve. Double block and bleed valves are especially useful in critical isolation or metering installations.

Ball valves

The inside component of a ball valve looks, not surprisingly, like a ball. As with the plug valve, a passageway is cut through the ball (fig. 4–20).

Fig. 4–20. Partially open ball valve

Ball valves rotate to open or close. When the ball is turned so the hole through the ball points along the pipe, fluids flow. When it points perpendicular to the flow, flow stops. Ball valves do not have forced mechanical seals. Rather, when the valve is closed, the differential pressure pushes the ball against the side of the valve body with the lowest pressure. In some cases the sealing surfaces (seats) on the body of the valve are spring-loaded against the ball, creating the seal. Ball and other valves can have a variety of sealing surfaces.

Ball valves are used for two different functions: isolation and control. Most often ball valves are either fully open or completely closed, serving the isolation function, in which case the passageway through the ball is cylindrical. When the inside diameter of the passageway is the same as the pipe's inside diameter, cleaning and inspection devices can pass through smoothly. The opposite is true of reduced port ball valves; the size of the opening through the valve is smaller than the diameter of the pipe. That makes passage of *pigs* (devices for cleaning, maintenance, or other tasks) difficult or impossible.

In some limited cases, ball valves serve a control function, and they are partially open. Fully open, the valve creates a minor pressure restriction. As the valve travels from open toward closed, the amount of pressure required to push gas through the valve increases, meaning the pressure on the downstream side of the valve is lower than the pressure on the upstream side of the valve. Passageways through ball valves performing the control function typically have special geometries to accomplish the desired pressure control.

Globe valves

When first manufactured, globe valve bodies were spherical, giving rise to the designation *globe*. Typically a circular disc moves up or down above a circular seat. Fully inserting the disc into the seat closes the valve. Lifting the disc from the seat allows flow (fig. 4–21).

In figure 4–21 the disc is labeled as a plug, but globe valves should not be confused with plug valves. Globe valves are used extensively to perform the control function for two reasons. First, as the disc moves up and down, the area of the opening varies linearly with the distance traveled, meaning the amount of pressure reduction achieved is somewhat more uniform throughout the travel of the disc than is possible with plug or ball valves. Second, flow is over the entire seat for any opening amount, meaning the seats will not erode as quickly, and if they erode, the erosion is relatively uniform around the seat.

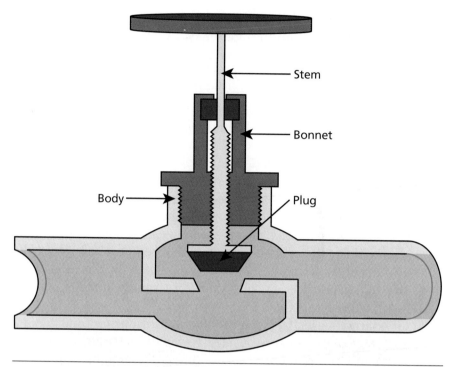

Fig. 4–21. Globe valve
Source: Released to the public domain by the creator.

As with many things, advantages for one application turn into disadvantages for others. Forcing the disc against the entire opening of the seat makes globe valves difficult to close at high pressures, so they are rarely used for isolation in high-pressure applications. They are used for isolation in distribution systems, particularly as angle valves where service lines are connected to mains.

Gate valves

Likely the oldest valve type, gate valves either consist of a slab of metal held in a frame that slides across the pipe, shutting off flow, or a slab of steel with a hole cut in it. When the hole is aligned with the pipe, the valve is open; when it is not, the valve is closed (fig. 4–22).

Tight seals are always an issue with valves, but high pressures are not a challenge for a slab gate valve. High differential pressure across the valve forces the slab against the seats, forming a tight seal. Sometimes valve seats are spring-loaded to help make a tight seal at low pressures.

Fig. 4–22. Slab gate valve

The gate valve slab is fastened to a long screw, called a *stem*, which in turn goes through a threaded handwheel. As the handwheel turns, the screw goes up or down, opening or closing the valve. Packing is tightly applied along the stem, keeping it from leaking. The main use of gate valves is to block flow completely. Gate valves were used in natural gas systems for years but have largely been replaced by plug and ball valves.

Check valves

Check valves come in multiple designs (swing, ball, and diaphragm) and allow flow in only one direction. Ball check valves contain a ball that is pushed out of the way by the force of the moving fluid. When the fluid stops moving, the check valve is pushed into a seat, either by a spring or by force of the reversing liquid, preventing further back or reversed flow (fig. 4–23).

Fig. 4–23. Ball check valve

Swing check valves, widely used in larger diameter lines, have a hinge-mounted flapper. The force of the moving fluid keeps the flapper pushed out of the way. When the fluid stops moving, the flapper falls closed and is held closed by the force of the fluid attempting to flow backward (fig. 4–24).

Fig. 4–24. Swing check valve

As long as the pressure upstream is lower than the pressure downstream, the valve remains forced closed by the pressure differential. When pressure builds on the upstream side sufficient to overcome the pressure on the downstream side, the valve opens.

Valve functions

Valves perform a variety of functions. The main functions include the following:

- *Isolate.* Blocks or shuts off flow completely.
- *One way or nonreturn.* Allows flow in one direction but closes when flow attempts to reverse.
- *Primary pressure control.* Modulates (opens or closes more or less) to control pressure and flow.
- *Monitor pressure control.* Installed in series with primary pressure control valves to serve as a backup. They "monitor" the pressure in the line and only function when the pressure rises above a predetermined level, indicating the primary pressure control valve is not performing its function.
- *Pressure relief.* Temporarily opens to reduce the pressure in closed lines or relieve pressure surges in flowing lines.
- *Excess flow.* Normally open but closes completely when the flow rate reaches a predetermined limit.
- *Seismic shut.* Isolates (shuts off) gas service when an earthquake of a sufficient magnitude occurs at a location.

Actuators

The devices that mount on valves to open or close them, or hold them partially open to control flow, are called *actuators*. They are often also called *operators*, as if that term was not used to refer to enough different things already. Actuators can be manual, pneumatic, electric, and hydraulic. *Manual operators* have been around since valves were invented. They consist of levers, wheels, and gears, and are turned by an operator standing next to the valve. The other forms of actuators can be operated either locally by a person standing next to the valve pushing a button or remotely in the same manner.

The plugs and balls in accordingly named valves have stems fastened to them. When the stem is rotated, the plug or ball rotates, opening or closing. Gate valves typically have a threaded stem connected to the gate. The threaded stem goes through a nut. When the nut is rotated, the stem goes up or down depending on the direction of rotation. Some globe valves have the same arrangement. The element in other globe valves moves up and down via a rod connected to a piston. Each valve needs an operator according to its movement mode: linear, quarter-turn, or multiturn. Check valves are operated by the flow and pressure in the line.

Pneumatic actuators

Pressure, either directly from the line or from an air compressor, provides motive force for *pneumatic actuators*. Diaphragm and piston are the most commonly used pneumatic actuators.

Diaphragm actuator. In its simplest form, the most commonly used industrial pneumatic actuator, it consists of a pressure containing vessel, a diaphragm, a spring, and a stem. The *diaphragm*, an elastomer membrane (sheet of stretchy material), is normally sandwiched between two plates (fig. 4–25).

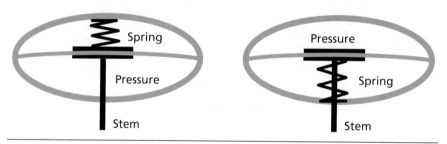

Fig. 4–25. Diaphragm actuator side view

The pressure side is connected to either natural gas pressure from the system or air pressure from station compressors. As pressure increases or decreases, the force on the diaphragm changes accordingly, compressing or decompressing the spring and thereby moving the stem up or down. The stem is connected either directly to the valve, causing the valve element to move up or down, or to some sort of gearing mechanism designed to translate linear motion to rotary motion. Figure 4–25 shows two designs. As pressure falls, the spring inside the actuator on the left moves the stem down, but the one on the right moves up. Total pressure loss means the stem on the left goes down as far as it can, but the one on the right goes up as far as it can. In other words, actuators can fail open or closed depending on system safety requirements.

Springs are selected based on operating pressure ranges. Most diaphragm actuators add an adjusting mechanism. When the mechanism is screwed in, the spring is compressed, requiring more force from the diaphragm (and more pressure from the system) to move the spring. The reverse is true for screwing the spring out (fig. 4–26).

Fig. 4–26. Diaphragm actuator top view. The adjusting mechanism is in the middle of the actuator.

Piston actuators. Like pistons in car engines, *piston actuators* consist of a piston driven back and forth in a cylinder. Air or natural gas pressure enters one end of the cylinder and drives the piston one direction, opening the valve. When it is time to close the valve, pressure is directed to the other end of the cylinder. Air or gas is then exhausted from the end in the direction of piston travel, decreasing the pressure on that side of the piston (fig. 4–27).

Sometimes pistons use hydraulic fluid (oil) introduced into the cylinder to drive the piston in the desired direction.

Fig. 4–27. Piston actuator powered by natural gas from the line

Electric actuators

Electric motor actuators are sometimes used to operate larger valves, and electrical solenoids are often used to operate smaller valves.

Electric motor actuator. In situations where natural gas is not available in sufficient pressure to power the actuator, or sometimes for safety reasons, electrical motors connected to the proper gearing arrangement open or close valves.

Solenoid actuator. Solenoids are electromagnets. When an electrical current is supplied to the solenoid, it moves a steel rod in one direction. When the current is turned off, the steel rod, and whatever is connected to it, moves back to its original position. Solenoid actuators often operate small valves that open or close to allow compressed air, natural gas, or hydraulic fluid to operate larger valves by directing pressure to one side of a diaphragm or a piston.

Meters

Before addressing meter types, a short discussion of natural gas measurement is in order. What is commonly measured is natural gas volume. What is commonly bought is natural gas energy. Thus, volume must be converted to energy.

$$\text{Energy content} = \text{Units of volume} \times \text{Energy content per unit}$$

Making the situation a bit more complex, only direct volume meters actually measure volume. Inference meters measure flow rate and calculate volume.

$$\text{Volume} = \text{Flow rate} \times \text{Cross-sectional area} \times \text{Time}$$

Further complicating natural gas measurement are compressibility, temperature, and pressure, as follows:

- One cubic foot of natural gas at 20 psi and 60°F contains more molecules (and hence more energy) than 1 cubic foot of natural gas at 10 psi and the same temperature.
- One cubic foot of natural gas at 10 psi and 60°F contains more molecules than 1 cubic foot of natural gas at 10 psi and 100°F.

The measurement process must convert the amount of natural gas contained in each cubic foot, at whatever its temperature and pressure, to the number of cubic feet that same gas would occupy at the agreed-upon (standard) temperature and pressure.

This correction can either be accomplished manually, automatically by devices located on the meter, or electronically by flow computers. Meters also normally have their own individual *meter factor*, a number peculiar to that specific meter used to correct for differences between what the meter says has gone through it and how much actually has gone through. Meter factors are important, established with provers, and vary with time as meters wear.

Direct volume meters

Also sometimes called *positive displacement meters*, the flow is separated into discrete segments, and the segments or chambers are counted as they go by. Adding the segments together determines total volume.

Diaphragm meters. The most commonly used low- to medium-volume delivery meter, a diaphragm meter consists of the following:
- A housing to contain pressure
- Elastomer diaphragms or bladders segmenting the meter into compartments of known volume
- Valves directing flow into or out of each compartment, accomplishing a rhythmic filling and emptying
- Small rods and levers connected to the bladders and valves causing the valves to open and close in proper sequence and the meter indicator to advance as each bladder or compartment empties
- An indicator (index) with arrows pointing to numbers indicating the amount of gas that has gone through the meter

Diaphragm meters come in different sizes for different capacities (fig. 4–28).

Fig. 4–28. Two different capacity diaphragm meters. The incoming service comes from the bottom of the picture and divides into a manifold with two meters currently connected. Note the additional connection point on the right, which does not currently have a meter installed.

Rotary meters. Precision-machined interlocking rotors form the basis of the *rotary meter*. They turn inside a case, displacing a specific volume of gas with each revolution (fig. 4–29).

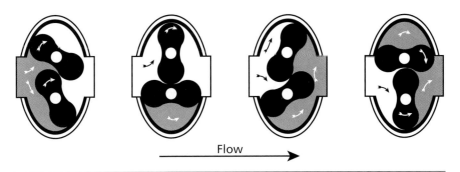

Fig. 4–29. Cross section of a rotary meter
Source: Dresser Installation, Operation and Maintenance Manual Series B3 ROOTS® Meters.

The operation of rotary meters is similar to a revolving door. They are typically used for larger volume and higher pressure metering than diaphragms.

Inferential meters

Rather than measure volume directly, *inferential meters* measure flow rate and calculate volume.

Turbine meters. Turbine meters began to receive recognition in the 1960s and have become well accepted as their technology has developed. They determine flow rate by measuring the speed of a bladed rotor suspended in the flow stream (fig. 4–30).

The rotational speed of the rotor is proportional to the flow rate. The volume flow rate is proportional to the cross-sectional area times the rotor's velocity, adjusted by the meter factor, and corrected for temperature and pressure.

Turbine meters do best when the flow rate is smooth and the properties of the fluid are consistent, that is, when density and viscosity do not vary. Turbine meters also like smooth, straight flow. That calls for a certain length of straight pipe prior to the gas entering the meter and after the gas exits the meter. *Straightening vanes*, essentially small pipes inside the larger diameter pipeline, can be installed if there is not enough room to install the straight pipe runs.

Fig. 4–30. Cutaway view of a turbine meter

Orifice meters. Bernoulli's principle concerning mass flow, velocity, and pressure is also the principle behind how orifice meters work. The meter consists of an orifice plate (steel plate with a round hole) of known size and configuration and sensors to measure line temperature, line pressure, and differential pressure from the front of the orifice plate to the back (fig. 4–31).

The hole in the orifice plate is smaller than the inside diameter of the line, so the gas velocity increases as it passes through the orifice plate. Since the gas is going faster, the pressure has to decrease. Using differential pressure across the orifice plate, line pressure, temperature, pipe diameter, and orifice plate configuration, gas quantities can be calculated either manually or electronically. Orifice meters are widely used for high-pressure, high-flow metering, such as at city gates.

Chapter 4 Components and Equipment 113

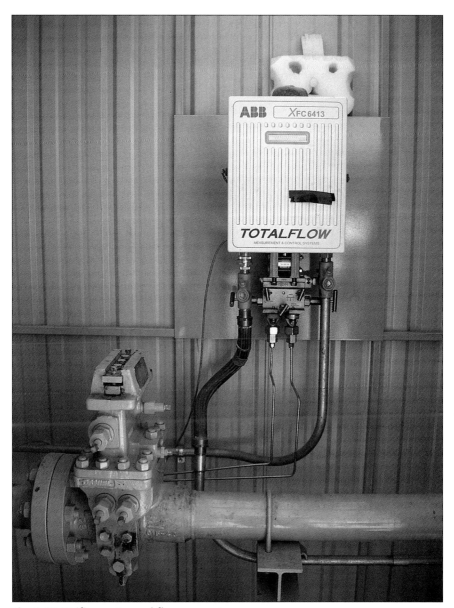

Fig. 4–31. Orifice meter and flow computer

Ultrasonic meters. Ultrasonic meters calculate flow rate from stream velocity, the cross-sectional area of the meter, and fluid density. They work by sending high-frequency sound pulses or waves from a transmitter on one side of the pipe to a receiver on the other side. The velocity of the flowing stream affects how long it takes for the pulse to travel across the pipe. Ultrasonic meters have several transmitters and receivers located around the meter. Each transmitter sends pulses frequently, and flow computers average the results to get accurate readings (fig. 4–32).

Fig. 4–32. Ultrasonic meter

Ultrasonic meters have several advantages over other meters. They are nonintrusive, meaning there are no obstructions to flow. They have no moving parts and cause no additional pressure loss. The precise timing of pulse measurements required to get an accurate reading is a challenge. Ultrasonic meters are gradually taking the place of orifice meters.

Coriolis meters. Gaspard-Gustave de Coriolis (1792–1843) discovered the Coriolis force, the principle on which these meters operate. This force, which has to do with angular momentum, also influences weather patterns, sea currents, and which way water turns as it flows down the toilet. For purposes of this text, it is sufficient to say that the mass, not the flow rate or velocity, of the fluid moving through a Coriolis meter acts on a vibrating tube, deflecting the tube. The amount of deflection is proportional to the amount of mass acting on the tube. Mass is then converted to volume. Coriolis meters are the most recently introduced meter type and are beginning to gain acceptance in the industry.

Reading meters

Meters used to be read manually by inspecting dials or digital displays (fig. 4–33).

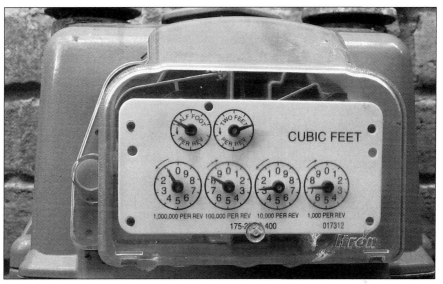

Fig. 4–33. Meter index on a standard diaphragm meter. The top two dials are used only during meter testing, not for reading the amount of gas purchased. The bottom four dials from left to right are: 0, 8, 2, 7. Accordingly, the meter reads 827, which is 827,000 cubic feet.

Many home meters, and nearly all large delivery and receipt meters, are now connected to electronic devices that periodically transmit the reading to a smart device either at the station or located somewhere in the neighborhood. This smart device then sends the readings on to a central location, where they are entered automatically into the LDC invoicing system.

Meter Provers

Meters are checked and calibrated at the factory before they are shipped. However, meter parts wear or become damaged, small amounts of dirt or debris accumulate in the system, or other metering conditions change over time. Meter inaccuracies are normally wrong in a consistent fashion; in other words, the mistakes are repeatable. Sometimes repeatable inaccuracies are remedied through physically adjusting the meter. More often, repeatable inaccuracies are remedied through the use of a meter factor calculated by dividing what the meter measures by what the prover measures.

As an extreme example, if the meter measures 100 standard cubic feet and the prover measures 110 standard cubic feet, during the same time, the meter factor is 1.1. Multiplying the meter reading of 100 by the meter factor of 1.1 yields the "correct" reading of 110. *Proving* is the process of determining the meter factor, and the devices used to prove meters are called *provers*. They are classified as either primary devices or secondary devices.

Primary proving devices

Primary proving devices such as sonic nozzles, bell provers, and volume provers work on slightly different principles. Nonetheless, each is calibrated to industry standards and measures either flow rate or flow amount.

Secondary proving devices

The master meter is the most often used device for proving natural gas meters. Master meters are called secondary devices since the master meter is proven using one of the primary proving devices. The master meter can then be connected in series with a field meter and used to prove the field meter. The flow goes through both meters, and the amount measured by the field meter over a period of time is compared to the amount measured by the master meter during the same period of time to arrive at the meter factor. Master meters are either mounted permanently at a location or are brought to the site as needed. Extensive industry standards deal with meters, provers, and measurement and proving practices to ensure accurate metering.

Odorant Skids

In its natural state, natural gas is odorless. The classic pungent smell most people associate with natural gas is actually a chemical injected into the gas stream at convenient locations. Odorant injection equipment, consisting of a storage tank and equipment to measure and inject the odorant, is normally placed at locations where the gas is also metered (fig. 4–34).

Fig. 4–34. Odorant injection skid located at a gas metering and pressure control facility

Rectifiers

Corrosion is caused by electrolysis, the same process that allows batteries to supply current to flashlights and other electrical and electronic devices. Corrosion in steel pipelines is commonly controlled by one of two methods. The first is through the use of coatings to stop current flow. The second is through installation of a rectifier and use of an electrical circuit to drive current in the desired direction, protecting the pipe (fig. 4–35).

Fig. 4–35. Rectifier

Rectifiers transform standard AC current to DC and reduce the voltage. The rectifier negative lead is connected to the pipeline. Completing the circuit involves connecting the positive lead to a metal structure called a *ground bed*, buried at a convenient location along the pipeline. Current then flows through the ground between the pipeline and ground bed. Over time, the ground bed rusts away and must be replaced. This process is called *cathodic protection* because, in electrolysis terms, the pipeline is the cathode and the ground bed is the anode.

Anodes

Burying magnesium or other types of metal along the line and connecting it to the pipe with a wire without a rectifier also causes current to flow, making the pipe the cathode. Anodes are supplied in convenient packages at various weights depending on the application (fig. 4–36).

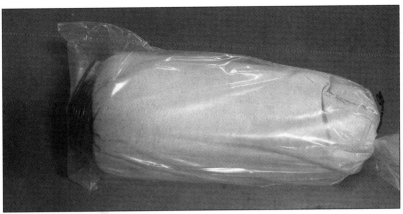
Fig. 4–36. Magnesium anode. Note the wire wound up on the left-hand side.

Anodes serve another purpose. They are fastened to the *tracer wire*, which is placed in the ditch with plastic pipe to facilitate finding the pipe after the pipe is buried. Tracer wires come above ground at each meter and at various other locations. Technicians fasten a device to the aboveground tracer wire for *energizing* or inducing an electrical current into the line. The anodes connected to the tracer wire along its route serve to draw current, increasing current flow, which then creates a magnetic field along the tracer wire. The strongest part of the magnetic field is directly over the buried pipe.

Storage

Natural gas demand varies greatly between seasons and throughout the day. The dead of winter requires more gas than a balmy spring day. When current production exceeds current demand, gas is put into storage. When current demand exceeds current production, gas is withdrawn. The primary storage types are as follows:

- *Depleted reservoirs.* Underground geologic features from which natural gas was previously produced.
- *Aquifers.* Impervious (nonporous) rock layers are located and wells are drilled through them. Gas is pushed down the well, displacing water. Hydrostatic pressure pushes the gas back out as needed.
- *Salt caverns.* Essentially caves created when salt is dissolved from naturally occurring underground salt formations.
- *Steel storage.* Natural gas is either liquefied or compressed and stored in steel vessels. The cost of steel storage limits its use.

Each storage type has its own maximum fill and withdrawal rate. Exceeding these rates can damage the structure. Consequently, fill and withdrawal rates are carefully monitored. Figure 4–37 from the US Energy Information Administration shows types and locations of natural gas storage facilities.

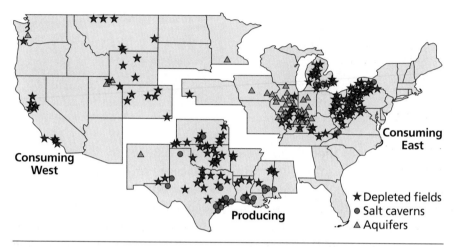

Fig. 4–37. US natural gas storage
Source: US Energy Information Administration.

Figure 4–38 from Gas Storage Europe shows, by country, the storage market size and the amount of that market serviced by underground storage.

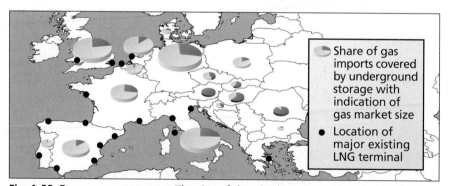

Fig. 4–38. European gas storage. The size of the pie chart represents the size of the market. The lighter slice of the pie shows the percentage of the market serviced by underground storage. Also shown are LNG facilities.
Source: Philipp Daniel Palada, "Gas Storage in Europe: Challenges and Outlook" (Presentation, Energy Community Workshop on Underground Gas Storage, GSE, May 28, 2014).

Another form of short-term storage is *line pack*. A large-diameter, high-pressure natural gas transmission line may start a cold winter day at 1,000 psi and drop to 750 psi or even lower as people awaken, turn up the thermostat, shower, and cook, withdrawing more molecules than are added. Line pack is not a long-term storage solution, but in the short-term, it is an important factor. Smaller diameter mains and service lines do not have the same swings in pressure but do provide a limited amount of supply cushion. Pressures must be carefully monitored and managed so they do not drop below critical operating levels.

Most LDCs own a limited amount of storage and depend on transmission and storage companies to meet their fluctuating demand.

Other Components

Pipe, coating, fittings, flanges, valves, actuators, meters, provers, odorant skids, rectifiers, anodes, and storage are the main or "mission critical" elements of a local distribution system. However, these are not the only components. There are also hundreds of instruments, sensors, transmitters, flow computers, strainers, scrubbers, knockouts, and other assorted components not discussed in detail here that enhance the efficiency, integrity, and safety of pipeline operations.

Summary

- Most distribution pipe is plastic, and most transmission pipe is steel.
- The six key properties of plastic pipe as listed by ASTM are "density, melt index, flexural modulus, tensile strength at yield, slow crack growth resistance, and hydrostatic strength classification."[3]
- Standard dimension ratio (SDR) is the ratio of the pipe outside diameter to the minimum wall thickness and indicates the pressure rating of the pipe.
- Specified minimum yield strength is a key property of steel pipe and is used in calculating maximum working pressure.
- Smaller (generally less than 6-inch diameter) plastic pipe is generally delivered to site spooled.
- Steel pipe is externally coated to prevent corrosion.

- Fusing plastic pipe lengths together involves heating the ends to melt the plastic and pushing the melted ends together, fusing them.
- Fusion bonded epoxy (FBE) is the predominant coating for steel pipe.
- Plastic pipe has been widely used since the 1950s, and extensive research is ongoing to understand the useful life of plastic pipe.
- Fittings are used to connect components or pipes and to allow changes in direction.
- Valves are classified by type and function.
- Valves isolate sections of the pipe from each other or throttle out pressure.
- Actuators move valve components, causing the valves to partially open or close, thereby controlling pressure and accordingly flow rate, or to open or close completely.
- Pneumatic actuators are the most commonly used actuators.
- Pressure from the gas stream is used to operate many actuators.
- The measurement process must convert the amount of natural gas contained in each cubic foot, at whatever its temperature and pressure, to the number of cubic feet that same gas would occupy at the agreed-upon (standard) temperature and pressure.
- Meters are either direct volume or inferential.
- Diaphragm and rotary meters are widely used in natural gas distribution. Orifice and ultrasonic are widely used for transmission.
- The master meter is the most often used device for proving natural gas meters. It is first calibrated by a primary prover.
- Rectifiers and ground beds are used to control corrosion of steel pipe.
- Natural gas storage may include use of depleted reservoirs, aquifers, caverns, steel storage, and even line pack, and provides important pressure and flow cushions.
- There are also hundreds of instruments, sensors, transmitters, flow computers, strainers, scrubbers, knockouts, and other assorted components.

Notes

1. American Gas Association, "Distribution and Transmission Miles of Pipeline," http://www.aga.org/Kc/analyses-and-statistics/statistics/annualstats/disttrans/Pages/default.aspx.
2. ASTM International, "Standard Specification for Polyethylene Plastics Pipe and Fittings Material," (ASTM International, 2014)
3. Ibid.

Natural Gas Transmission Line Operations

We do everything that needs doing, including cleaning the bathrooms.

—Second-generation natural gas field operations supervisor

At 2 AM on Christmas Day, three controllers sit at their consoles on the 24th floor of an office tower located in Houston, Texas, gazing intently at the Weather Channel. The screen displays a massive cold front sweeping into the northeastern United States. As expected by the gas supply staff when they built the schedules the previous day, record lows are forecast. Turning to their consoles, the controllers nod knowingly. With efficient clicks of their mouse devices, they drill down through successive computer screens. Starting at the origination station southwest of Houston and working methodically up the line toward New York City, they click away, bringing on more compression. Additional natural gas is pushed into the line from storage located along the Gulf Coast, packing the line. The controllers realize that in just a few hours, children will awaken to look under the Christmas tree. Thermostats will ramp up, demanding gas for furnaces and stoves, and Christmas lights will come on, demanding electricity from gas-fired power plants. Figure 5–1 shows compressor station locations on the US natural gas pipeline network.

As the line packs from storage on the downstream end, the controllers turn their attention to consumption points farther upstream, examining several screens in succession (fig. 5–2).

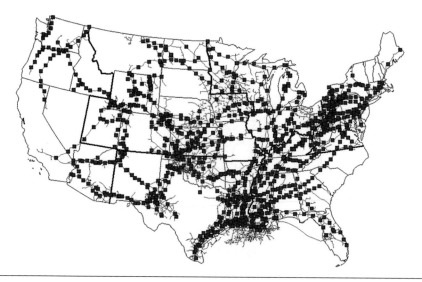

Fig. 5–1. Natural gas compressor stations located on the US natural gas pipeline network. In 2006 there were more than 1,200 operational natural gas compressor stations located on the US interstate natural gas pipeline network, containing more than 4,700 individual compressor units, averaging nearly 3,600 horsepower per unit.
Source: US Energy Information Agency.

Fig. 5–2. Typical gas control console. These screens are part of the typical human machine interface (HMI).

Directions developed by schedulers guide the controllers, but they temper the schedules with years of experience, clicking their mouse devices as they go. Occasionally they pick up the phone to speak with technicians called out for emergency repairs or to check with controllers running connecting pipelines. Soon the controllers will be clicking again, causing gas currently flowing into storage to reverse direction and flow out, satisfying demands. Figure 5–3 shows the locations of storage fields.

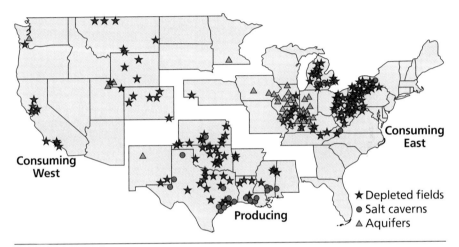

Fig. 5–3. Natural gas storage fields. Owners/operators of underground storage facilities typically are (1) interstate pipeline companies, (2) intrastate pipeline companies, (3) local distribution companies (LDCs), and (4) independent storage service providers. About 120 entities currently operate the nearly 400 active underground storage facilities in the lower 48 states.
Source: US Energy Information Agency.

Six PM approaches, and the off-going controllers brief the next shift prior to heading home. The next shift's controllers examine their line configurations (pressures, rates, horsepower, and compressors in operation). Making minor adjustments, starting a compressor here and shutting down one there, the controllers adjust line operations to suit their operating preferences. The pipeline controller's job is much like that of an air traffic controller. It is routine most of the time but is interspersed with moments requiring prompt action to avoid upsets or even disasters. Day in and day out, 24/7, controllers ensure gas flows reliably and safely, supported by schedulers, technicians, engineers, and a myriad of others.

This chapter deals with natural gas transmission pipelines. It covers local field operations and central control room operations, extends the hydraulic concepts to natural gas, and finishes with the less-than-pleasant subject of abnormal operations. There is one important note regarding natural gas safety. Most people think natural gas has a distinctive odor, but it is naturally odorless. The distinctively sharp odor comes from ethyl- or methyl-mercaptan. One of these odorants is added as the gas approaches the point of consumption to help the public detect a leak.

Natural Gas Lines: A Brief Review

As discussed in chapter 1, natural gas lines are comprised of gathering, transmission, and local distribution lines. Figure 5–4 shows the natural gas value chain.

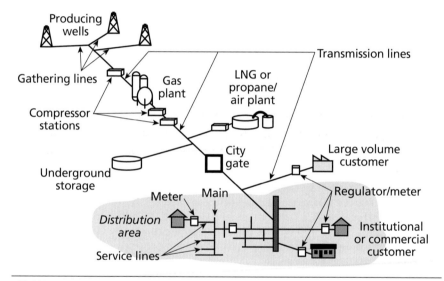

Fig. 5–4. Natural gas value chain

On the upstream end, natural gas gathering lines extend from natural gas production to gas processing plants and transmission pipelines. Natural gas transmission pipelines transport for shippers who sell to large users (factories, schools, malls, or electrical generation plants, for example), local distribution companies, and gas marketers. Of course the large users or local distribution companies (LDCs) may ship the gas on transmission pipelines themselves, purchasing it directly from producers before it enters the transmission line. Some companies own production, gathering, transmission, and local distribution lines through different subsidiaries. They may even own gas processing plants or have a trading company, buying and selling gas between related and nonrelated entities.

On the downstream side, natural gas LDCs sell to many of the same customers as those served by transmission pipelines and to individual homes and businesses. They receive from natural gas transmission pipelines and LNG gasification plants, and also from their own storage or from other storage.

The distinction between natural gas transmission lines and local distribution lines has more to do with history, regulations, traditional customers, line size, and operating pressures than hydraulics and operating principles. Natural gas, regardless of the pipeline or country it is in, obeys the same universal laws of physics.

While the focus of this book is on natural gas local distribution pipelines, a few words regarding natural gas processing, gathering, and transmission are helpful precursors to the remainder of the text.

Gas processing plants

Some gas wells produce *dry gas*, which is essentially methane with very little heavier hydrocarbons or water vapor. If it has no other contaminants, the gas can enter transmission lines without additional processing. However, in most cases, raw (unprocessed) gas contains as much as 20% heavier molecules such as ethane, propane, and butane (fig. 5–5).

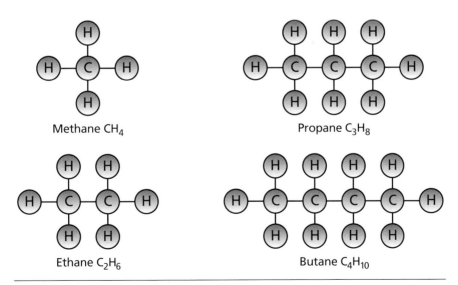

Fig. 5–5. Hydrocarbon molecules. The primary constituent of natural gas is methane, although some of the larger hydrocarbon molecules may also be present in small amounts.

Even after the initial separation in the production field, most raw gas still contains water vapors and other contaminants such as hydrogen sulfide and carbon dioxide. All of these, as well as the heavier hydrocarbon molecules, must be removed before natural gas (composed of nearly pure methane) is injected into the transmission line (fig. 5–6).

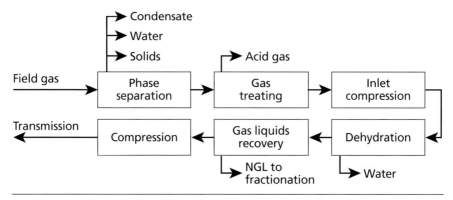

Fig. 5–6. Simplified gas processing plant flow. Gas processing plants may also contain fractionation equipment that further divides the NGL stream into its constituent components, such as ethane, propane, butane, and heavier (larger) molecules.

Water vapor

Gas processing plants (often called simply *gas plants*) use one of two methods, the names of which sound a lot alike, to *dehydrate* or remove water vapors from the gas stream.

Absorption employs liquids, like glycol, which attract water molecules. Natural gas passes through the liquid, which grabs on to the water vapor. Gas is lighter than the liquid, so the liquid sinks to the bottom of the stream. It is drawn off and sent to a boiler for water vapor removal, after which the liquid is sent back for another pass.

Adsorption uses solid materials having an affinity for water. Usually employing at least two towers packed with absorbent materials, the gas stream is directed to one tower at a time. When the material in the first tower becomes saturated, the stream is switched to the next tower while the first is regenerated. The towers cycle one after the other, first adsorbing and then regenerating.

Heavier molecules

Following water removal, other heavier hydrocarbons are extracted. One of two processes, absorption or refrigeration, is employed. Absorption, like water absorption, makes use of oil that attracts larger molecules from the gas stream. Before it contacts and absorbs the ethane, propane, butane, and other heavier molecules, the oil is called *lean oil*. The name changes to *rich oil* when it is saturated with the natural gas liquids. Drawn from the bottom of the absorption tower, rich oil is fed to lean oil stills, where the lower boiling point molecules are driven from the oil, returning it to lean status.

Assuming the same pressure, methane turns into a liquid at lower temperatures than do ethane, butane, and propane. Refrigeration simply cools the gas stream to below about –120°F, liquefying all but the methane molecules. One popular refrigeration process, called *turbo expansion*, rapidly expands the gas stream across a turbine, dropping the temperature. However, recall the concept of conservation of mass and energy. If the gas cools, where does the heat go? It becomes mechanical energy to turn the turbine. Energy passed to the turbine by the gas as it expands provides some of the mechanical energy needed to recompress the methane for injection into the transmission line.

Fractionation

Separated from the methane, the residual stream still contains ethane, propane, and butane, along with some pentanes and heavier molecules. In successive towers, the stream will next undergo fractionation, passing through a deethanizer, a depropanizer, and a debutanizer. One type of molecule at a time is boiled off in each tower. The residual stream, pentanes and heavier, is sold into the fuels or chemicals market.

There are four reasons for removing the water and heavier hydrocarbons; three are operational and one is economic.

Slug flow. Heavier hydrocarbon molecules and water have lower dew points than natural gas (methane). If left in the natural gas stream, they can cool and condense into liquids as they travel along the pipeline. These liquids collect in low spots, restricting and eventually completely blocking gas flow. Pressure upstream of the blockage builds, finally forcing the slug of liquid over the next hill and on toward the next low point. This results in *slug flow* and inefficient pipeline operation (fig. 5–7).

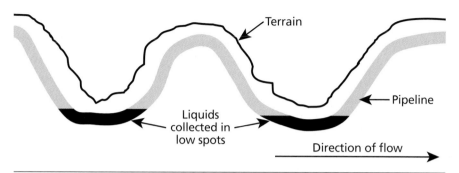

Fig. 5–7. Slug flow. Liquids collect at low spots, temporarily blocking flow. Pressure builds upstream of the low spot, eventually pushing the liquid over the upstream hill, causing a surge. Worse, water can collect and freeze, stopping flow entirely.

Internal corrosion. Water collecting in low spots is one cause of transmission pipeline internal corrosion. Catastrophic consequences can occur if the water is allowed to remain in the line (fig. 5–8).

Fig. 5–8. Crater left by natural gas explosion. Internal corrosion caused the natural gas leak near Carlsbad, New Mexico, which resulted in an explosion causing 12 fatalities.
Source: US National Transportation Safety Board.

Even after processing, some liquids may fall out of the natural gas stream later, so lines are designed with intentional low spots, called *drips*, or special towers, called slug catchers, and filters to capture and safely remove accumulations of liquids before they cause problems.

Btu content. For operating and safety reasons, natural gas appliances such as furnaces, water heaters, washers, and dryers are designed to operate in certain Btu ranges. The standard is usually about 1,050 Btu per standard cubic foot. Ethane, propane, butane, and natural gasoline have successively higher Btu content than this. The gas may have to be processed to extract higher Btu components, particularly the butane and natural gasoline. The energy content of ethane is closer to methane and can be left in the gas.

Economic value. In addition to the practical operating reasons for gas plants, there is an economic reason. Under most market conditions, ethane, propane, butane, and natural gasoline are worth more than methane, making gas plant construction and separation of these hydrocarbons commercially viable. At times, the conditions reverse and ethane or propane is left in the gas stream, and sometimes it is even injected. This can cause conflict between natural gas sellers wanting top dollar for their Btu levels and natural gas pipeline operators wanting consistent natural gas quality. Natural gas pipeline controllers closely monitor incoming gas streams for energy content and heavier hydrocarbon content. High volume receipt points may have gas chromatographs to monitor gas quality (fig. 5–9).

12:00:00 MONTH/DATE/YEAR	
CURRENT	STREAM 1
BTU	1036.7
SCV	0.5871
%NIT	0.336
%CO2	0.894
%METHANE	95.61
%ETHANE	2.308
%PROPANE	0.5
% I-BUTANE	0.113
% N-BUTANE	0.115
%NEO PENTANE	0
% I-PENTANE	0.041
% N-PENTANE	0.027
%C6 PLUS	0.054
% HEXANE	0.027
%HEPTANE	0.027
NEOC5 + IC5	0.041
UNNORMALIZED TOTAL	100.6801
HOURLY	
BTU	1036
SGV	0.5863
%NIT	0.331
%CO2	0.89
DAILY	
BTU	1030.1
SGV	0.5831
%NIT	0.329
%CO2	0.91

Fig. 5–9. Typical readings from a gas chromatograph

Gas gathering lines

Natural gas starts its journey to market from underground reservoirs, coming to the surface through individual wells (fig. 5-10).

Fig. 5–10. Natural gas well. So-called Christmas trees are located atop wells. Oilfield workers adjust the various valves and gauges to control pressures and flows. Steps are a safety feature to allow easier access.

Gas from each well, often mixed with oil, water, and contaminants, flows from individual wells through flow lines to central processing areas. There, separators and treaters divide gas from oil and water and remove some impurities. Next the gas moves into gathering lines for its journey to transmission lines. These lines are usually between 4 and 10 inches in diameter, with the actual size determined by the amount of gas required.

Sometimes the gas flows all the way to transmission lines driven only by well pressure. Normally gas is measured as it enters gathering lines, and often small compressors are installed to move the gas along (fig. 5-11).

Fig. 5–11. Gathering meters and compressor. The building in the foreground contains an orifice meter and flow computing equipment powered by the solar cell located at the top. In the background the compressor adds pressure to the gas.

Additional lines may connect (tie in) along the route until the gathering line eventually arrives at the gas plant or transmission line. A familiar gathering line analogy is neighborhood streets that progressively lead to larger city thoroughfares and eventually to the freeway system.

Transmission lines

Transmission lines range from a few miles to thousands of miles long. They serve as the autobahn for the natural gas industry, hauling massive amounts of energy from gas plants, LNG facilities, other pipelines, and storage fields to homes, schools, office buildings, commercial establishments, manufacturing plants, and electricity generation plants. Transmission lines may be as small as 6 inches and as large as 48 inches or more in diameter. Some transport natural gas from just a few origination points to a limited number of destinations, but many have several hundred delivery and receipt points.

The United States and Russia, according to the CIA *World Factbook*, together contain more than one-half of the almost 800,000 miles of natural gas transmission lines in existence. Figure 5–12 shows a more detailed breakdown.

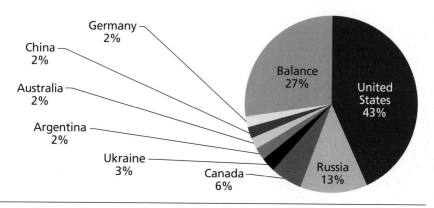

Fig. 5–12. Percentage of natural gas transmission mileage by country
Source: Compiled by Miesner, LLC; data from CIA, *The World Factbook*.

Receipts from liquefied natural gas (LNG) facilities

In the last few decades, LNG has become an important part of the natural gas value chain in Europe and East Asia. LNG imports into the United States have been cyclical depending on market conditions. LNG and its impact on natural gas pipelines merit mention.

What is LNG? It is methane, the same colorless, odorless hydrocarbon molecule in current natural gas pipelines, but it has been cooled to –260°F (or –162°C) in an LNG plant, where it turns from a gas to a liquid. In liquid form, natural gas occupies about 1/600th of the volume it had in its gaseous state. The smaller volume makes it economically feasible to load it on specially designed tankers for ocean transportation from producing to consuming countries.

LNG import terminals temporarily store the LNG until it is warmed back up to normal pipeline temperatures and delivered into transmission lines. At first blush, handling LNG in a natural gas line would seem to present no liquids problems. After all, when LNG is created at its origin, the gas has to be cooled to –260°F to get it into the liquid state. In the process, long before that temperature is reached, the ethane, propane, butane, and heavier molecules liquefy and could easily have been removed from the natural gas stream. But LNG often comes from *stranded gas*, a term that implies the gas production is located in some remote place, far from energy markets. If that is true, the market for natural gas liquids may be equally isolated. The owner of the LNG operation may be interested in leaving the natural gas liquids in the LNG as a means of transporting them cheaply to a market or increasing its Btu content. Some LNG can even be spiked with extra propane and butane because the destination market pays a premium for the liquids. In this case, the liquids-laden LNG may pose the same liquids problems covered earlier. For this reason, LNG specifications and quality are receiving considerable attention.

Pipeline hubs

The ability to receive from and deliver to other pipelines is important for supply flexibility and reliability. Over time, regional *pipeline hubs* have developed, locations where a number of pipelines interconnect. Hubs are important as physical interconnects, allowing gas to move efficiently from one pipeline to another. However, they are also important as clearing points for gas trading. Examples of hubs include the National Balancing Point (NBP) in the United Kingdom, Zeebrugge in Belgium, Baumgarten in Austria, and the Henry Hub in the United States (fig. 5–13).

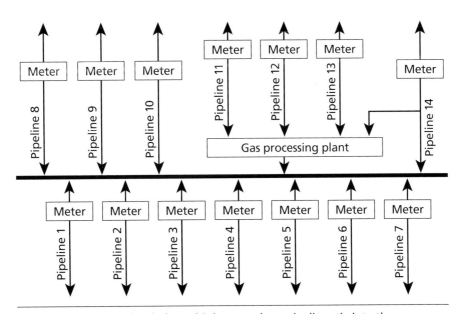

Fig. 5–13. Typical pipeline hub. Multiple natural gas pipelines tie into the same "header." Gas is metered in and metered out, allowing transfer between multiple parties.
Source: Pipeline Knowledge & Development, *Oil and Gas Pipeline Fundamentals.*

From field operations to central control room operations

In the 1860s, the Oil Transportation Association provided the first real-time (or at least near real-time), end-to-end communications system along a pipeline. They built a telegraph along the route of a two-inch, wrought iron pipeline over a six-mile track from an oil field to the railroad station at Oil Creek, Pennsylvania. Over the telegraph line they communicated the complex bookkeeping of multiple crude oil shippers. This innovation began the trend to central control room operations.

As technology developed, the telephone complemented the telegraph and motors replaced the manual handles on valves. Dispatchers sat in central control rooms and communicated by phone (or telegraph). Still, operations continued to be labor-intensive. Operators located in the field told the dispatchers what was happening, and the dispatchers told the operators when to open or close valves, start and stop compressors and engines, read meters, and perform other operations. Dispatchers were orchestra conductors, telling the operators what to do but depending on the operators to do it.

As automation took over, central control rooms became the nerve centers of the pipelines, and the roles of dispatchers and operators merged into the controller's job. Much of field operations moved to the central control room, where controllers could read conditions from displays, decide what to do, and send out electronic commands to remote compressors, valves, and the like. Central control rooms depend heavily on communications technology and electronic controls to do the work formerly done manually by operators and over the telephone by dispatchers. And from down the hall, the schedulers feed them the plans to implement.

While many field operating jobs have moved into the central control room, pipeline maintenance continues to be carried out mostly along the pipeline route by technicians and pipeline service companies. The administrative and regulatory function of operations is growing, requiring more people to keep up with the regulations and to document compliance. A few more words differentiating operations from maintenance are in order.

- Operations cause the pipeline to function such that it performs its intended purpose.
- Maintenance keeps the pipeline in operating condition at its current capacity.
- Field operations are those conducted along or near the pipeline's route.

- Some maintenance activities such as planning are conducted away from the pipeline, but most maintenance activities are conducted on the pipeline or along the route.
- Control room operations are conducted at limited numbers of locations remote from the pipeline.
- Operations and maintenance are often performed by the same people.

The balance of this chapter's focus is operations. Chapter 9 addresses maintenance.

Natural gas field operations

Field operators are still critical to the success of natural gas pipelines. They measure gas (including the vital component of quality testing) and interact with landowners, contractors, and city and state officials. They make decisions regarding compression and other operating variables and serve as valuable "eyes and ears" for central control room controllers. The same person often contributes to operating, maintenance, administrative, and regulatory activities. In the words of one natural gas field operating supervisor, "We do everything that needs doing, including cleaning the bathrooms."

Measurement. "The meter is the cash register," says the pipeline operator. Thus one of the most important roles of field operators is measurement. Recall the difference between liquids and gases. Both conform to the shape of their container, but gas expands to fill the container and more easily escapes if it is not contained. As a result, gas measurement differs from liquid measurement. The most common gas measurement technique is the orifice meter, although other metering techniques such as turbine meters, ultrasonic meters, and Coriolis meters are also used. One note: ultrasonic meters are slowly gaining popularity over orifice meters.

Orifice meters are based on Bernoulli's principle (chapter 3). They consist of an orifice plate of known size and configuration and sensors to measure line temperature, line pressure, and the difference in pressure (*differential pressure*) upstream and downstream of the orifice plate. The orifice plate contains a hole smaller than the inside diameter of the line, so the gas velocity increases as it passes through the orifice plate. Since the gas is going faster, the pressure has to decrease. Using differential pressure across the orifice plate, line pressure, temperature, pipe diameter, and orifice plate configuration, gas quantities can be calculated either manually or by computer. Temperatures, line pressures, and differential pressures can fluctuate significantly, even over short time periods. Flow computers have largely replaced paper charts and manual calculations, particularly in high-volume locations (fig. 5–14).

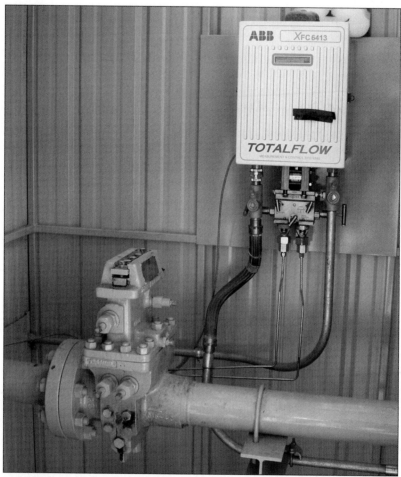

Fig. 5–14. Orifice meter with flow computer. The small diameter tubes from the orifice meter on the lower left connect to the flow computer on the upper right, allowing it to sense pressure changes from computed flow.

Besides quantity, quality is important, and the raw gas stream is sampled to determine the distribution of methane, ethane, propane, butane, heavier hydrocarbons, and contaminants in the stream. For low flow applications, automatic samplers are normally used. Samples are taken at prescribed intervals and stored in pressurized containers (fig. 5–15).

Chapter 5 Natural Gas Transmission Line Operations 141

Fig. 5–15. Gas sampler. A small amount of gas is withdrawn through the probe installed in the line and is stored in the small pressure container in the foreground right.

The pressure container is changed on a regular basis. In the lab, the gas sample is analyzed for energy content (in British thermal units). Contaminants such as carbon dioxide, oxygen, and nitrogen are also determined and accounted for in the payment calculations.

When contaminant concentrations reach certain levels, the pipeline company may refuse to continue accepting the natural gas, shutting down production.

There is one final note regarding metering. To ensure accuracy of measurements, the pressure and temperature gauges are calibrated on a predetermined schedule and also whenever problems are suspected. Calibration is carried out by field technicians who compare pressure and temperature readings of the field instruments to reference instruments with known accuracy, or against deadweight testers that produce a known pressure. Adjustments, either manual or electronic, reestablish field instrument accuracy as needed.

Managing stakeholder relations. Field operators and supervisors are the main contact point for stakeholders along the route. They work with landowners, understanding and addressing their concerns, help train local emergency response personnel, and advise local officials regarding siting issues. They also work with excavators, helping them determine pipeline locations important to safe digging. While the management of stakeholder relations is not actually an operating task, it increasingly is emerging as one of the key responsibilities of field operators and supervisors. Time invested building trust with landowners and local officials pays big dividends when operating problems arise or if permits are needed to expand or even simply maintain the lines.

Right-of-way clearing, which is keeping a strip of land immediately over the pipeline free of trees and other large vegetation, is a maintenance function. It is, however, mentioned briefly in this stakeholder section because landowners, understandably, are concerned when others enter their property. Regulations (and good operating practices) require regular line patrols. They look for construction activity, line exposures, and other potential hazards. Typically, right-of-way patrol is conducted from the air by airplanes or helicopters, making clear visibility critical. Regular right-of-way maintenance seldom creates concern. However, clearing neglected rights-of-way, especially ones containing mature and scenic trees, is a source of friction.

Compressor station operations. *Origination compressor stations* are compressor stations located at the beginning of the line. Figure 5–16 shows the outside of an origination station, and figure 5–17 is a partial view of the engines and compressors inside the compressor building.

Chapter 5 Natural Gas Transmission Line Operations 143

Fig. 5–16. Origination gas compressor station. Several gathering lines from onshore and one from offshore enter this mainline origination compression station. The natural gas is aggregated into a common suction header. A combination of the seven reciprocating compressors, driven by natural gas–fired reciprocating engines, adds pressure to the gas before it is discharged into a 30-inch natural gas mainline.

Fig. 5–17. Inside of an origination station with positive displacement compressors. From a catwalk about 10 feet from the floor in the compressor station shown in figure 5–16, five cylinders of the V-10 engine are visible.

Farther along the line, *booster stations* add more energy, giving the gas a boost as it travels along its journey (fig. 5–18).

Fig. 5–18. Inside of a booster station with centrifugal compressors. Electric motors are often used in stations to meet emission standards. In addition, the station may be insulated for noise reduction. The electric motor is on the left, with a variable speed drive and the compressor connected to it in that order.

Engines require cooling and lubrication. Cars have radiators and motor oil providing these functions. Compressor stations have cooling and lubricating systems for the same purposes. Compressor cooling systems sometimes cool discharge gas also. (Recall that compression adds energy to press molecules closer together, increasing both temperature and pressure.) Control equipment monitors station discharge temperatures to ensure they are not too high. An important function of station operators is dealing with this ancillary equipment to keep the station operating.

Natural gas delivery operations

One entity's receipt is another's delivery, so it should not be surprising that delivery operations have the same operating practices and equipment as receipt operations.

Gas interchange points. Multiple pipelines connecting to each other in an area form hubs. When only two connect, that location is normally called an *interconnection point* (fig. 5–19).

Fig. 5–19. Gas interconnect. Valves and actuators are located in the back right of the picture. An ultrasonic meter is located in the middle of the picture just past the crossover. Monitoring equipment in the metal control building sends signals to the central control rooms of the two gas transmission lines that interconnect at this point.

City gates. At city or town gates, transmission lines deliver natural gas to LDCs (fig. 5-20). At that point, pressure regulators reduce pressures down to those allowed in the distribution systems. There are also metering facilities, quality monitors, and odorization facilities. Heaters may be needed to warm the gas that cools as a result of the pressure reduction at the station.

Fig. 5–20. City gate. Gas enters the station through the large pipeline in the foreground. Note the control building in the background and the radio tower to the left of it. This tower transmits volumes, pressures, quality, and other operating information to the central control room.

Both LDCs and the transmission companies often have their own sets of meters at the city gate. Sometimes they agree to share meters, and the information from one metering station is transmitted back to both control rooms. Several transmission companies may supply an LDC through different city gates, and large markets have a number of city gates.

Flow control and pressure control. Deliveries to LDCs at city gates or other large users such as power generation or industrial users are either flow controlled or pressure controlled. Flow control limits customers to a maximum withdrawal rate, usually expressed in cubic feet or dekatherms per hour. Pressure control allows customers to withdraw what they need as long as the delivery pressure remains above a minimum level. Even with flow control, pressure control may also be in effect to guard against low pressures on the line.

Central Control Rooms

Staffed 24 hours a day, seven days a week, central control rooms are the pipeline's nerve center. Schedulers work with shippers, develop operating plans, and hand them over to controllers for implementation. Computer screens are the controllers' eyes and telephones their ears as they diligently survey monitors and optimize line operations (fig. 5–21).

Fig. 5–21. Typical gas control console. Each monitor contains a different screen with the information the controller needs to understand line operations and take actions to control the pipeline.

The functions of the central control room include the following:
- Planning and scheduling operations
- Forecasting available capacity
- Receiving shipper nominations
- Developing pump, receipt, and delivery schedules
- Making adjustments to the plans as needed
- Communicating the plans and adjustments
- Monitoring and controlling the pipeline
- Understanding the operating scenario
- Operating equipment (compressors, valves, motors, meters, etc.)
- Monitoring line and equipment parameters
- Making routine adjustments
- Communicating line operations
- Responding to calls from third parties
- Responding to and correcting abnormal operations
- Detecting abnormal operations and conditions
- Responding to abnormal and emergency situations
- Keeping all parties informed as needed
- Dealing with nonroutine communications
- Maintaining the control equipment
- Planning and scheduling operations

Scheduling

The scheduler's job seems simple enough: balance supply and demand. Schedulers receive estimates, normally called *nominations*, from shippers, and arrange a schedule that matches supply with demand (fig. 5–22).

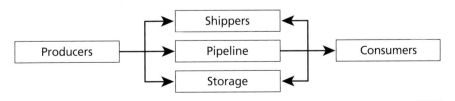

Fig. 5–22. Pipeline control room work flow

However, pipelines have from several to hundreds of receipt and delivery points. Demand can cycle as much as 100% from summer to winter and from nighttime to daytime. Often the day's dawn brings unexpected logistical constraints on top of planned slowdowns. Power disruptions caused by storms, production curtailments because of equipment problems, and surprise equipment breakdowns are just a few examples of situations dealt with by schedulers daily. Other internal and external factors complicate the work flow (fig. 5–23).

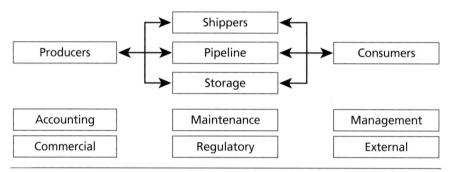

Fig. 5–23. Pipeline control room work flow with added constraints. Unexpected events and external factors impinge on the central control work flow, demanding accommodation.

Natural gas pipelines, transmission as well as distribution, are closely interlinked, mandating cooperation. Problems on one pipeline quickly spread to others. In North America, the North American Energy Standards Board (NAESB), a voluntary industry group, developed a common set of scheduling and nomination standards. These standards were then approved by the US Federal Energy Regulatory Commission (FERC) and other regulatory bodies.

The protocol has four cycles or iterations for each day. Each cycle requires the pipeline scheduler to address or carry out the following:
- Receive shipper nominations
- Validate these nominations with connecting pipelines
- Build an ideal schedule based on nominations
- Balance the ideal schedule with the real world
- Maintenance
- Other constraints
- Contract provisions

- Communicate the trial schedule to shippers
- Get shipper feedback
- Make adjustments as needed

In the United States and Canada, the four cycles begin at the following times:

- Cycle 1 (Timely)   11:30 AM of the prior day
- Cycle 2 (Evening)—6:00 PM of the prior day
- Cycle 3 (Intraday 1)—10:00 AM of the day
- Cycle 4 (Intraday 2)—5:00 PM of the day

Each cycle goes through the same steps. Some pipelines are even moving toward hourly balancing.

Europe, through the European Union and others, recognizes the value of liberalizing gas markets. Some standards exist to facilitate gas trading and scheduling, and shippers move gas on and off balancing zones. Shippers must have entry capacity to enter and exit capacity to exit the zone. However, countries can still have different gas day start times, quality allowances, shipper codes, and nomination requirements. Some industry groups are working hard to drive more standardization, interconnections, hubs, and balancing points. Others are resisting the move. Gazprom, as the largest player, has a big say in what happens. Asia, with less-developed infrastructure, recognizes the need for infrastructure and the standards to make it work.

Schedulers handle fluctuations by cutting back supply, filling or withdrawing from storage, or packing and unpacking the line. *Packing* involves putting more into the line than is taken out, while *unpacking* is naturally the reverse. Line pack can swing as much as 30% and provides a valuable operating cushion.

Natural gas pipelines normally receive from and deliver to each other directly without going through storage. Receiving pipelines (particularly LDCs) often use the pressure from the delivering pipeline and have no compressors of their own. Low pressure from the delivering pipeline affects the receiving pipeline immediately. High pressure at the receiving pipeline causes the delivering pipeline problems. Balancing supply, demand, and pressures among hundreds of receipt and delivery points requires constant attention. Demand is sensitive to time of day, day of the week, weather, and even sunshine and cloudiness. Schedulers adjust the supplies to meet overall demand on a daily, weekly, and monthly basis. Controllers, discussed later in this chapter, adjust ongoing line operations, accommodating needs as they happen.

Each receipt and delivery point must be tracked to account for ownership. This task normally falls to an accounting group, but they work closely with the schedulers, controllers, and measurement people to ensure accurate accounting. Companies employ individual practices, programs, and procedures to effectively schedule and account for natural gas moved on their systems.

Successful schedulers do not mind dealing with minutiae and are often called on to make quick decisions. Informing a producer they are about to be shut down for quality reasons or explaining to shippers why they must move their gas out of storage even though it means a distressed sale are normal job requirements. Their backgrounds are varied. Some, desiring a day job, have moved over from the controller ranks. Others have accounting, financial, or field supervisory, operator, or technician backgrounds. Whatever the background, patience and communication skills are important scheduler traits.

Monitoring and controlling the pipeline

In the words of one natural gas controller, "Our job is to safely manage line operations, ensuring pressures are low enough so suppliers can get their gas into the line but high enough customers have enough pressure to withdraw their needs, all at the lowest cost possible." Natural gas controllers manage pressures and line pack aided by SCADA and control systems covered in chapter 10. Three of the primary SCADA tools natural gas controllers use are the following:

- System maps with control points
- Pressure and flow summary
- Station graphics

System maps. These screens show the system map and pressures and flow rates by location. From these the controller can see how the system is balanced (fig. 5–24).

Pressure and flow summary page. Graphics are helpful in some cases, but when a lot of information needs to fit on one screen, text is sometimes the best approach, as shown in this pressure and flow summary page (fig. 5–25).

Chapter 5 Natural Gas Transmission Line Operations 151

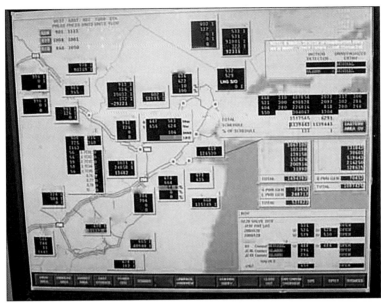

Fig. 5–24. System map with pressures. Typically only the large delivery points are shown on the map. In this case the power plant delivery connections are shown.

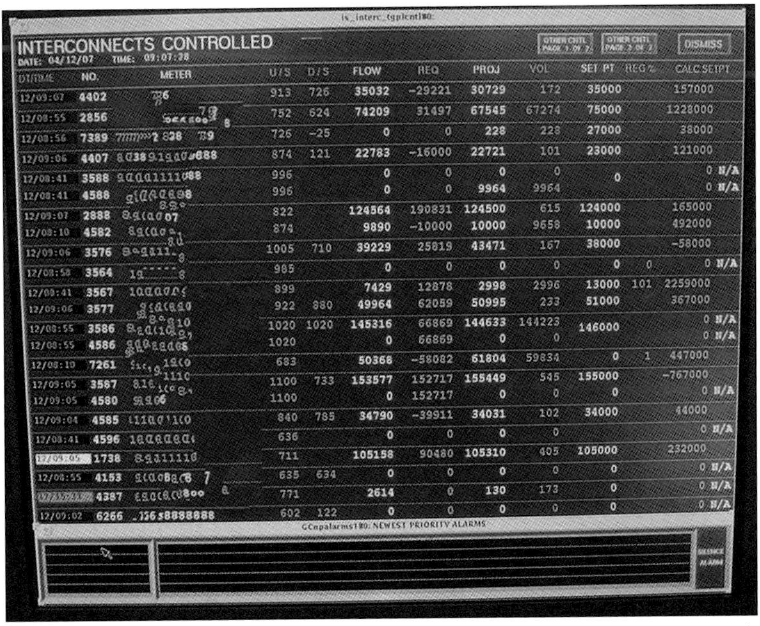

Fig. 5–25. Pressure and flow summary page. Flow rate and pressure are shown for each connection point, even the small ones. Experienced controllers know exactly where to look on this page of dense text to find the number they need.

Station graphics. The station graphics page is one of the key pages allowing controllers to issue control moves, that is, to make something happen to the line. Each little symbol represents a piece of equipment at this meter station. Controllers click on points with their mouse devices, drilling down for more detailed information. When they decide to open a valve, they click on the valve symbol to bring up a dialogue box. In the box they click on Open or Close, such as the information about a particular group of delivery locations (fig. 5–26).

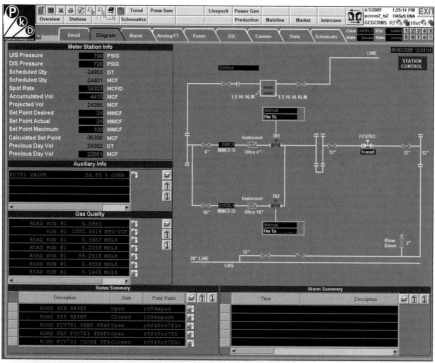

Fig. 5–26. Meter station graphics page. The main 20-inch gas pipeline is shown at the bottom of the screen. Graphics show the major piping, valves, and meters associated with the station. Buttons along the top of the page allow easy access to additional screens containing more information.

Optimizing line operation

Pipeline controllers study discharge pressures, flow rates, and line pack to optimize line operations. They accomplish this by bringing more or less compression online or by sending natural gas to storage or withdrawing it, always keeping every location on the line below its maximum allowable operating pressure (MAOP).

For some pipelines, controllers decide which compressors to start or stop. On other lines, controllers tell the control system how much station horsepower is desired, and the station computer decides which compressors to run, based on a priority list programmed into the computer by local station operators.

Routine operations

Natural gas pipeline operators want pipeline pressures as high as safely allowable, since higher pressures translate to less friction loss per mile per unit moved, and more efficient operations. Producers, and others delivering into the pipeline, want pipeline pressures as low as possible, enabling them to input maximum gas quantities with as little compression on their side of the delivery valve as possible, saving them money. LDCs and others receiving gas are generally aligned with the natural gas pipelines, wanting high delivery pressures, enabling them to operate with as little compression as possible.

Pipeline controllers communicate with and balance the needs of these different groups, making frequent decisions regarding compressor configuration and line pressures. They also manage the day-to-day input and output upsets. When a major gas plant unexpectedly shuts down, curtailing supply, controllers examine the situation and adjust line operations to compensate. They may decide to withdraw gas from storage, stop delivering into storage, or simply take some compression off-line on the supply end, allowing the line to unpack. If a cold front arrives earlier than anticipated, increasing demand, controllers have to compensate again, using a combination of the same mechanisms of storage, compression, or line pack.

Sometimes controllers access computer modeling programs to assist with decisions about compressor selection and other line operations. Generally, for large transmission lines, the models have limited reliably because of the hundreds of receipt and delivery points, each with its own set of constraints and constantly changing conditions. Computer models are improving but at most natural gas pipelines, controller experience and intuition are still vital to safe and efficient operation.

Quality control

Natural gas controllers play an important quality role. They monitor receipts to ensure gas quality is within specifications. At major receipt points, gas chromatographs monitor the line, displaying line quality as shown in chapter 4 (fig. 4–9).

Control systems sound alarms when properties of the incoming stream exceed preset limits (high or low), and operators take corrective action. While it is a last resort, they may shut down receipt points that violate pipeline specifications.

Abnormal operations

A chapter on operations would not be complete without addressing the topic of abnormal operations. The concept is not always ominous; unplanned or unexpected abnormal operations can cover a wide range of situations. Some of them are given here:

- Unintended closures of valves, shutdowns, or activation or deactivation of any device (motor, valve, etc.)
- Increase or decrease in pressure or flow rate outside normal operating limits
- Loss of data in the control center
- Operation of a safety device
- Inability for the control center to control equipment

Unintended closures, activations, or deactivations. According to Murphy's law, sometimes valves will fail or close inadvertently, or other devices will operate incorrectly when no other problem exists. However, a valve closing unexpectedly can also indicate something terrible has happened, such as a leak. Whether false alarm or real problem, controllers must be trained to overreact, not underreact. The cause must be determined quickly. Continuing to operate the line or restarting the line is risky without understanding and correcting the problem.

Increase or decrease in pressure or flow rate outside normal operating limits. Unexpected pressure or flow rate changes often indicate a leak. If a break occurs in the section of line close to and downstream of a compressor station, the line friction loss drops quickly because the gas is traveling a shorter distance. Flow rate in that section of line goes up. The compressor station likely shuts down because of low suction pressure into the station or high flow rate out.

If the break happens relatively far from the compressor station, close to the downstream station, the pressure loss at the compressor station does not drop as quickly. However, the compressors downstream, at the next station, are starved since less (or no) gas is coming to them. The station goes down, again on low suction pressure. In both cases, low suction pressure indicates a major line break.

Loss of communications. Communications, supplying accurate data about line operations and allowing controllers to remotely control the lines, is one of the keys to central control rooms. Without the ability to see or control, the controller is relegated to the dispatcher role, calling to ask questions and telling field operators what to do. That might have worked in the old days when each station was manned, but it may not be possible today due to limited field operating forces. Detailed plans tell controllers what to do in the event of communications loss and involve quickly reestablishing communications or shutting down the line.

Operation of any safety device. Safety devices are installed as the fail-safe protection, just like the backup parachute carried by skydivers. Ideally, they never function. But when they do, it means something went wrong. When a skydiver's primary chute fails, the skydiver does not just say, "Oh well, my backup saved me!" and nonchalantly go on to the next jump. Instead, extensive investigations are conducted to determine why the primary chute did not open. Corrective actions are taken to ensure it does not happen again. If operation of safety devices becomes normal and expected, controllers are desensitized to what could go wrong. They could begin to depend on safety devices to handle problems, always a dangerous approach. Operators constantly remind themselves that safety device activation is not normal operation.

Inability for the control center to control equipment. Akin to loss of communications, but more likely caused by a malfunction at the local station or in the central control room, it carries the same consequences. SCADA and control designers build in redundancy and safeguards, but nothing is ever 100% fail-safe, so controllers must have contingency plans and understand the need to overreact versus underreact.

Precautions during maintenance activities

Maintenance activities call for special care. Controllers must know the timing and type of maintenance so they can operate the line appropriately. Examples of such activities include the following:
- Running internal inspection devices
- Lowering, reconditioning, or welding on the line while in service
- Performing pressure tests

These situations are carefully planned and monitored and are a normal but infrequent part of maintenance. Controllers are usually briefed concerning response to these situations before they occur so the details are fresh in their minds. Additional information is covered in chapter 9.

Natural gas releases

Small amounts of natural gas escaping from pipelines are essentially invisible. Large leaks are found quickly because the rapidly decompressing gas causes a crater and makes a roaring noise. For large leaks, the control system alarms a pressure drop and controllers shut down the line and close valves along the route, allowing only the natural gas between the two closest valves (one upstream and one downstream of the leak) to escape.

Smaller leaks are more difficult to detect. Chapter 8 covers leaks and leak management, discussing the challenges line pack presents to finding small natural gas leaks in high-volume transmission lines. Leaking natural gas kills the surrounding vegetation, a clear sign of potential trouble to the alert operator patrolling the line on the ground, or the line flier patrolling from the air. Ground patrols also employ gas "sniffers" designed to detect small levels of natural gas. Emerging technologies detect leaking hydrocarbons from airplanes and satellite.

Operator training

The job of a controller is a bit like air traffic controller and pilot built into one. It requires a unique combination of skills. These include the ability to understand complex systems stretching over many miles where one action influences another, along with the ability to quickly respond to upset and emergency situations.

On-the-job training plays a large role in controller training. It is effective in teaching operators how to handle the routine operations of their particular line. However, since abnormal operations happen infrequently, on-the-job training is not as useful in preparing operators to handle these situations. Instead, simulations, drills, discussions, and playback of previous problems play a more effective role in preparing operators to respond. Most companies have extensive training programs for their controllers before they are ever allowed to "solo."

Emergency response

This final section discusses what to do in the event of an emergency. Natural gas is not toxic, but it is highly flammable if mixed with oxygen. Natural gas pipeline operators develop extensive emergency response plans detailing the response command structure, what to do in the event of an emergency, and how to do it. They also conduct training and mock drills involving their own people and local, state, and federal emergency response officials.

In general the immediate response plans call for the following:

- Determine the necessary steps to protect life and property, including isolating the danger zone and evacuating people if necessary.
- Isolate the affected section of pipe.
- Shut off the flow of gas. (Block valves are installed along the line for this purpose.)
- Allow the remaining gas, which is lighter than air, to dissipate.
- Prevent ignition of the escaping gas. (If the gas does ignite, it is normally allowed to burn itself out while the surrounding areas are protected from secondary ignition.)

Interestingly, pure natural gas will not ignite. It needs oxygen to support combustion. At natural gas concentrations of less than 5%, there is not enough gas to ignite. Greater than 15%, there is not enough oxygen.

After the emergency has passed, the pipeline company assesses property damage resulting from the incident. They also preserve the integrity of the site and work closely with local, state, and federal agencies to determine the cause of the incident and appropriate remedial measures.

Summary

- Natural gas transmission pipelines receive natural gas from gathering systems, gas plants, other transmission lines, storage, and LNG regasification plants.
- They deliver to local distribution lines (through city gates), large customers, other transmission lines, storage, and sometimes LNG liquefaction plants.
- Movements through transmission lines vary seasonally.
- Flow rates and line pack fluctuate during each 24-hour period.
- Storage includes long-term seasonal storage and intraday storage.
- Gas plants remove heavier molecules, water vapor, and other contaminants.
- Operations cause the pipeline to function such that it performs its intended purpose.

- Maintenance keeps the pipeline in operating condition at its current capacity.
- Often the same people perform operations and maintenance functions.
- Managing stakeholder relations is commanding increased attention.
- Central control rooms are the nerve centers of the pipelines.
- Schedulers balance supply and demand through an iterative scheduling process.
- Controllers manage pressures, flow rates, and line pack on an hour-to-hour and minute-to-minute basis.
- Controllers require the divergent skill sets of thoughtful optimization and quick actions.
- Like crying wolf, repeated false alarms desensitize controllers.
- Mock emergency response drills sharpen skills and teach responders to work together.

chapter 6

Local Distribution Pipeline Operations

Kites rise highest against the wind, not with it.

—Winston Churchill

    At 7:30 AM Friday, the supervisor of measurement and pressure control for an LDC serving a city of nearly 750,000 people calls an emergency meeting at the service center. Several months ago the natural gas transmission pipeline company serving as the LDC's primary gas supply contracted for a routine internal inspection of the lateral line serving the city. Late Thursday the inspection contractor had informed the transmission company of an anomaly requiring immediate repair that was discovered during preliminary analysis of the pipeline inspection data. Thus on Saturday the transmission company will shut down the line for repairs, removing the primary gas supply to the east side of the city.

    Another transmission company is connected to the same east-side city gate, but if it has problems, a large part of the city, including the power plant for the university, will be completely out of gas. Carefully the supervisor and technicians plan their course of action. They decide which tasks already planned for the day should be delayed and which must go on as planned. Then one technician heads for the city gate to double-check all meters, pressure control, and other equipment. Another coordinates plans by phone with both transmission pipeline companies. A third volunteers to come in on Saturday to "babysit" the city gate until the transmission line is back in service. The supervisor adjourns the meeting and heads back to the office, mentally rejuggling the day's work on the way.

Around 10 AM, the supervisor climbs into a truck and heads out to check with the technicians. The technician at the city gate is actively rebuilding a pressure controller and has called in an additional technician to check meter calibrations and the communication system back to the central control room. The supervisor has an idea and checks with the manager by phone. They agree that physically patrolling the backup transmission line's route Saturday is an appropriate precaution. The supervisor phones one of the technicians and asks him to patrol the line Saturday to keep an eye out for excavation work along the line that could cause problems. The technician agrees and calls to rearrange the family's weekend plans.

Such is the life of field operators. Moving between planned and unplanned activities, they constantly reshuffle priorities, professional as well as personal, to keep the gas flowing safely, reliably, and efficiently.

Local Distribution Pipelines: An Overview

Local distribution systems are composed of city gates, mains, service lines, associated equipment, and occasionally storage or other means of providing supplementary gas. These distribution systems pick up at the city gate where transmission lines leave off and end at the customer delivery meter. US Government statistics report there are nearly 2.2 million miles of natural gas local distribution lines in the US, comprised of about 1.3 million miles of distribution mains and about 0.9 million miles of service lines.[1] In *Statistics 2008*, Eurogas reported the EU-27 member countries plus Switzerland and Turkey had a little over 2 million kilometers of natural gas pipelines. Of those, about 1.8 million kilometers (1.1 million miles) were distribution pipeline (fig. 6–1).

In subsequent reports, Eurogas discontinued reporting transmission and distribution kilometers separately, but it is reasonable to believe that the distribution number remains between 1.8 and 1.9 million kilometers.

LDC ownership broadly falls into three categories: private, public, and state owned. Private ownership comes in the form of corporations (stock) or partnerships (units). Individuals, pension plans, mutual funds, and other traditional investors own the stock or units. Public ownership, sometimes called *municipal ownership*, means cities, towns, or countries own the LDC. *State-owned* is a term applied to centrally planned rather than free enterprise companies.

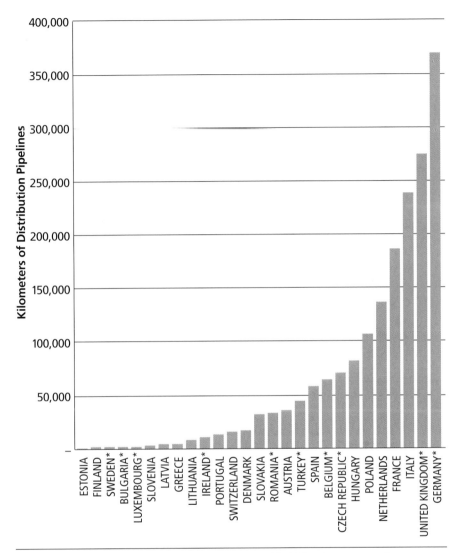

Fig. 6–1. Distribution mileage by country. An asterisk (*) by the country denotes an estimate.
Source: Compiled by Pipeline Knowledge & Development from *Statistics 2008*, Eurogas, http://www.eurogas.org/uploads/media/Statistics_2008_01.01.08.pdf.

City gate stations

Sometimes called *city gates, city border stations, town stations,* or *town taps,* these connection points between gas transmission pipelines and distribution lines reduce pressure, measure gas, provide one more cleaning function, and inject the smelly chemical most people associate with natural gas. Pressure reduction results in expansion, which takes energy, thus cooling the gas. Chapter 3 explained the cooling effect of gas expansion (fig. 6–2).

Fig. 6–2. Pressure reduction valve. The pressure reduction valve is on the right. Note the frost on the pipe immediately to the left of the valve.

Some city gates include heaters to warm the gas back up to normal operating temperatures. Heaters minimize the risk of impurities, such as water, freezing due to the temperature drop and thus creating operational issues for the operator.

Mains

Mains are roughly separated into two categories based on their operating pressures. High-pressure mains, sometimes also called *feeder mains*, *supply mains*, *interstation mains*, and *intermediate pressure mains*, operate at various pressures, usually between about 300 and 60 psi. *Distribution mains*, also called *low-pressure mains*, operate at lower pressures, usually less than 60 psi. The name and pressure range have to do with local regulations and company practice rather than any precise operating condition. The American Gas Association defines them as:

    a. High pressure. *A system which operates at a pressure higher than the standard service pressure delivered to the customer; thus, a pressure regulator is required on each service to control pressure delivered to the customer. Sometimes this is referred to as medium pressure.*

    b. Low pressure or utilization pressure. *A system in which the gas pressure in the mains and service lines is substantially the same as that delivered to the customers' appliances; ordinarily a pressure regulator is not required on individual service lines.*[2]

For purposes of this book, the terms *high-pressure mains* and *distribution mains* are used to differentiate the two. In the United States, diameters typically range from 2 to 12 inches, with less than 1% greater than 12 inches in diameter (fig. 6–3).

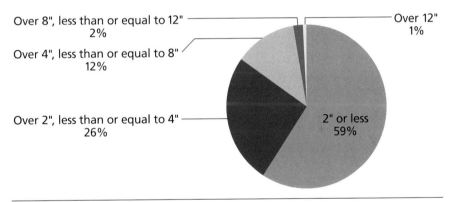

Fig. 6–3. Diameter ranges of mains in the United States. Approximately 85% of US distribution mains are 4-inch diameter or smaller.
Source: Developed by for this report by Cheryl Trench, Allegro Energy Consulting, from PHMSA Form F 7100.1-1 data.

Most distribution systems have several interconnected mains operating at progressively lower pressures as they near the customer.

Early mains were constructed primarily of cast iron pipe. However, cast iron is brittle and may crack or fail due to soil stresses. The bell and spigot joining technique used with cast iron pipe sometimes pulled apart, allowing leaks. Over the years LDCs have conducted extensive programs aimed at replacing or rehabilitating their cast iron mains.

New mains are constructed of plastic or steel. Polyethylene (PE) plastic pipe, introduced in the 1960s, comprises more than 90% of the approximately 40,000 miles of new mains constructed each year in the United States (fig. 6–4).[3]

Fig. 6–4. Installing a plastic main. This 6-inch diameter plastic main installed to serve a new subdivision is fused together. Note the tracer wire in the left foreground. Part of the main has been lowered in and is ready for backfilling.

District regulators

Recall from chapter 3 the concept of pressure reduction caused by friction. Similar to city gate pressure reduction, district regulators introduce constrictions that increase friction and reduce pressure. District regulators are located along mains as needed to control pressures. Some are located above ground, but they are often buried in underground vaults (fig. 6–5).

Ensuring pressure reduction is critical, so often a second, redundant regulator referred to as a *monitor* is located either upstream or downstream of the district regulator. As one would expect, it monitors the first regulator, taking over if the first one fails. Sometimes rather than being buried, they are installed above ground (fig. 6–6).

Chapter 6 Local Distribution Pipeline Operations   165

Fig. 6–5. District pressure regulator installed in a vault. The pressure reduction valve is in the middle of the picture between the two manually operated valves. The small pipe coming out of the top connects to the main farther downstream (in the direction of flow). It feeds back information regarding downstream pressure, causing the pressure regulator to open or close.

Fig. 6–6. Aboveground district pressure regulator with backup monitor valve. The pressure reduction valve in the background, partially obscured by the yucca plant, serves as a backup to the primary located in the foreground. Note the barriers installed between the regulator station and the road.

Service lines

Distribution mains generally operate at pressures lower than 60 psi and carry gas to service lines. Service lines tap into distribution mains and carry gas to meters at customers' facilities. Larger use customers have larger diameter lines to carry more gas with the same amount of pressure loss per unit. Pressure regulators are normally located at a customer's facility to reduce pressures to the level furnaces and other gas-burning devices are designed to handle. Normal delivery pressures are about 0.5 psi but may vary depending on local regulations and the particular system. Some distribution mains operate at pressures low enough that pressure regulators are not needed at customer locations. Most service lines are less than 2 inches in diameter, and less than 1% are larger than 2 inches in diameter (fig. 6–7).

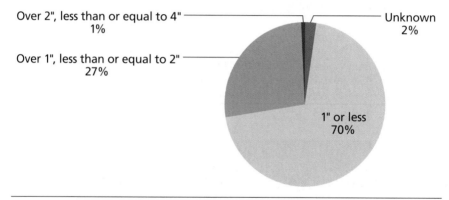

Fig. 6–7. Diameter ranges of service line in the United States
Source: Developed by Cheryl Trench, Allegro Energy Consulting from PHMSA Form F 7100.1-1 data.

Sometimes, such as in the case of new subdivisions, service line taps are installed during original construction. At other times they are installed later by simply digging up the line and installing a tap. Of course this process can be quite costly in congested urban areas, so the industry is working hard to develop solutions that minimize the installation footprint. In addition to taps on distribution mains, sometimes high-pressure mains are tapped for new services if a distribution main is not located nearby.

Meters

Most people hardly notice the familiar natural gas meter attached to the side of their house (fig. 6–8).

Fig. 6–8. A newly installed meter. The meter is in the center of the picture and the pressure regulator is on the left. Immediately below the regulator is a shutoff valve for turning off the gas. Note the screwed fittings.

The gas company usually owns facilities up to and including the meter. House piping is installed by a plumber working for the homeowner or home builder. After the house piping is installed, the construction department, or a specially trained contractor, installs the service line connecting the main to the meter. Then the gas company's service department installs the meter, regulator, and associated valves and fittings. In many cases the meter installation includes an automatic shutoff valve called an *excess flow valve*, normally located at the beginning of the service line or at the property line of the customer. In the event flow reaches preset high flow rates, consistent with a leak, the excess flow valve closes automatically.

Farm taps

Sometimes single locations such as farms, ranches, factories, and other remote users desire natural gas service, giving rise to so-called farm taps. These small installations consisting of a valve, pressure regulator, pressure relief, and sometimes one or more meters are usually direct connections to transmission lines. Sometimes meters are located at the tap site and sometimes at the customer's location, which can be as far away as several miles from the tap. Over time farm taps may grow to serve several sites as population spreads out and may even grow into city gates as more equipment is added to handle the increased demand. Farm taps are hybrids between gate stations, service lines, and customer meters, serving all three functions.

Supplemental gas

When considering supplemental gas, it is important to note that there are least three different industry definitions. One definition of *supplemental gas* is gas used seasonally and daily to balance supply with demand, regardless of its type. Another refers to gas used in specific locations to maintain service to an area. A third definition is gas used to supplement traditional natural gas supply, such as liquefied natural gas, LPG-Air, synthetic natural gas, landfill gas, biomass gas, coal gasification, and unconventional gas. This book uses the first two definitions. Storage, peak-shaving facilities, and line pack are important tools to keep gas flowing smoothly.

Storage. In the days of manufactured gas, gas was normally stored in aboveground gas holders (fig. 6–9).

Due to their small volumes and low pressures, gas holders are now obsolete, replaced by underground storage from depleted oil and gas reservoirs, salt dome caverns, and aquifers, or in limited cases by aboveground pressurized or refrigerated storage (table 6–1).

Fig. 6–9. An early gas holder. As gas entered the holder, pressure pushed the top up, and a series of interlocking rings or courses telescoped up, forming the side walls. *Courtesy:* National Fuel Gas Company.

Table 6–1. Storage types, numbers, capacities, and withdrawal rates in the United States

	# of Sites	Working Gas Capacity (Bcf)	Daily Withdrawal Capability (MMcf)
Reservoirs	327	3,532	64,698
Salt Domes	29	169	15,404
Aquifers	43	397	8,358
Totals	399	4,098	88,460

Source: Compiled by Pipeline Knowledge & Development based on US Energy Information Administration data.

As mentioned earlier, natural gas demand varies both seasonally and intraday. Transmission line pack can accommodate many small changes, but the large demand swings are usually handled with storage. When it comes to storage, there are several important considerations:

- How much natural gas can the storage hold?
- How high can storage pressures be safely raised without risking the integrity of the storage?
- How fast can gas be withdrawn from it?
- How fast can gas be reinjected?
- Are sufficient mineral rights leased to minimize the risk of others drilling into the storage and claiming native gas production?
- What impurities, if any, will be picked up by the gas stream in storage that must be removed during withdrawal?

Natural gas storage is currently receiving considerable attention on an international level. Market participants strive to balance daily and seasonal supply and also try to capture market opportunities.

Peak shaving. According to the American Gas Association, peak shaving is "the use of fuels and equipment to generate or manufacture gas to supplement the normal supply of pipeline gas during periods of extremely high demand."[4] One common peak-shaving facility is an LNG plant. LNG is stored close to where natural gas is needed and is regasified to supplement existing supply. Another peak-shaving technique involves mixing LPG, normally in the form of propane, with air until the Btu content of the mixture mirrors that of natural gas, at which point the mixture is injected into the supply grid. This can be done either at the town border station or at a point in town where increased pipeline pressures are desirable.

Maintaining service

A common guideline when designing natural gas supply grids is, "Designed-in redundancy of feed." In other words, supply locations are from more than one point. Sometimes it is relatively easy and does not cost much extra to follow this rule (fig. 6–10).

Adding the extra main in figure 6–10 provided two-way feeds to the houses, but this subdivision still only has one line supplying it. One-way feed sometimes happens, particularly as new subdivisions are built farther out on the grid. As the market develops, the LDC engineers will look for ways to convert to two-way feeds.

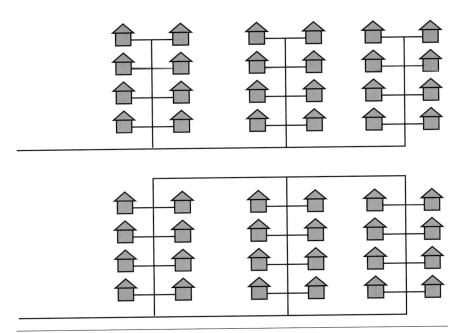

Fig. 6–10. Supply redundancy. The top layout feeds each house from only one source. Adding a main to form a loop as shown on the bottom layout feeds each row of houses from both ends, improving supply reliability.

In the case of one-way feed, shutting down the incoming feed when it requires maintenance means the subdivision would be out of service. Portable compressed natural gas (CNG) or LNG tanks are sometimes used to solve this problem (fig. 6–11).

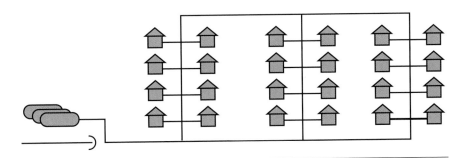

Fig. 6–11. Schematic of temporary CNG or LNG supply

A temporary supply of CNG or LNG is positioned between the area needing repair and the customers and is tied into the main. Then the main is shut down upstream of the connection point. The main can be evacuated and repaired while the customers are fed by the temporary supply. Once repairs are completed, the line is put back into service and the portable plant moves to another location or back to the service center for later use.

This is not to say customers are never out of gas, but LDC employees work hard to minimize any disruptions, especially ones that can be avoided with planning. In addition to maintaining customer relations, minimizing outages reduces the costs of checking customer piping and relighting pilot lights following an outage.

City gates, mains, regulators, service lines, meters, and storage all play critical roles when it comes to providing safe, reliable, and affordable natural gas service. Next is the most important topic concerning pipeline operations: the people who make it all work.

Functions and Tasks

Natural gas distribution pipelines obey the same laws of physics as natural gas transmission lines, meaning many of the functions and tasks are similar. However, gas distribution lines are generally smaller in diameter and operate at lower pressures. By their nature, natural gas distribution pipelines are located in populated areas, such as cities and towns, whereas transmission lines are largely rural. These three differences—diameter, pressure, and population density—mean some functions are quite different from those performed by transmission company employees.

All companies have organizational structures. However, the organizational structures often differ even between companies in the same business, influenced by personalities, capabilities, local preferences, and other factors. Figure 6–12 is a function chart, not an organization chart. It shows at a high level the tasks required to run an LDC and groups them into five main headings for convenience.

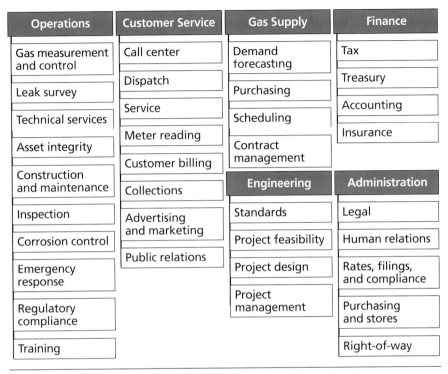

Fig. 6–12. LDC function chart. Organization charts may look quite different from function charts.

Operations

Each and every gas furnace, stove, oven, water heater, or other appliance must receive gas within its specified pressure and Btu range. If the pressure or Btu content drops too low, the flame goes out. High pressures and high Btu content bring safety risks. People expect perfection from natural gas operations: continued gas flow 24/7, at the proper pressure and energy content, and without any discernible interruption. Anything less falls short of the mark.

Gas measurement and pressure control

Natural gas demand constantly changes during the day, driven by the time of day, day of week, season, ambient temperature, precipitation, and a myriad of other factors. Changes in flow velocity and the associated friction losses can cause pressure variations. During peak demand the pressure at the city gate must still be sufficient to push the gas to the most distant home or business and burn smoothly. Gas measurement and pressure control are all about ensuring enough (but not too much) pressure, while accurately measuring the amount and quality of incoming and outgoing gas.

Some LDCs employ centralized control rooms to monitor pressures and control flows over their entire service area. Others take a more decentralized approach, with localized control rooms across the service area. Whatever the approach, gas measurement and control works closely with gas supply. Gas supply ensures gas is available at the city gate. Gas measurement and control makes sure it gets to the ultimate customer.

Accomplishing this task involves manually or remotely opening and closing valves and adjusting pressure regulators. Manually operated pressure regulators ensure system pressures remain below preset levels. They sense the downstream pressure. As it approaches a preset upper limit, they close, causing friction and dissipating pressure. The opposite happens as downstream pressure falls, until the regulator is completely open. Manual regulators, once they are adjusted, operate automatically within a preset range. Gas control operators or field operators open and close remotely operated regulators to fine-tune system pressures. Local technicians adjust manual regulators as needed. They also inspect, maintain, and repair regulators and meters as required (fig. 6–13).

Fig. 6–13. Technician repairing a pressure control valve. Flow is from left to right. The technician is adjusting the pilot spring on a pressure control valve.

Finally, there are measurement and pressure control technicians who calibrate incoming meters to ensure accuracy. There are also technicians responsible for the odorant injection facilities. They make sure that mercaptan, the smelly substance that gives otherwise odorless natural gas its distinctive smell, is added according to federal regulations to ensure safety.

Leak survey

On a regular schedule and systematic basis, technicians survey every mile of local distribution systems in search of leaks. More on leaks is included later in this text.

Technical services

Technical services is an engineering function that focuses on existing assets. Sometimes it reports to the operations manager and sometimes the engineering manager. Regardless, the key is how well it supports operations.

Serving as important training and reference tools, operating guides and equipment manuals are normally developed, updated, and maintained by technical services personnel. Keeping these guides and manuals current is no small task. They are quite detailed and taken together can fill several yards of shelf space or terabytes of hard drive space. Technical services personnel do not do all the work themselves, though. They are assisted by subject matter experts (SMEs) from other departments who share knowledge they have gained over the years.

Pipeline system maps are a far cry from the familiar foldout road maps. Geospatial technology is often used and is essentially the same global positioning system (GPS) technology many people use in their cars. However, pipeline maps go way beyond satellite positioning technology. Many gas utilities are building geographic information systems (GIS), electronic visual databases that provide their map systems with layers of information.

Pipeline companies purchase base GIS packages containing commercially available layers and then add system-specific layers to them. These layers include information such as the pipeline route, materials of construction, pipeline diameter, and wall thickness, as well as the location of valves, regulators, meters, and other pipeline features. Repair and replacement history and landowner names and information can even be stored in these databases. Maintenance and inspection history of valves, fittings, odorizers, and other appurtenances can be kept in a GIS system as well. All this information is tied to locations along the line, providing critical information at the click of mouse. But just like home inventories, checkbooks, and other personal records, keeping this much data current and up-to-date is a huge task.

GIS information facilitates system planning, such as forecasting demand, analyzing the resultant flow rates and pressures, and developing ways to reinforce the system. System planners feed GIS data about lengths, diameters, materials of construction, regulator capacities, and sometimes other variables into computer models. These models utilize customer consumption history to calculate peak flow rates, estimate friction loss, and model pressure drop along the system under peak-day conditions. They point out potential bottlenecks and constraint areas. Engineers investigate a variety of solutions to the bottlenecks, including increasing pressure, replacing mains with larger diameter pipe, adding loops, adding receipt points, or sometimes a combination of options. Each scenario is analyzed with the model and checked against construction and operating cost to arrive at the optimum solution.

A variety of other functions may be included under the auspices of technical services. These include performance benchmarking and improvement, failure analysis, records management, design and materials standards, technical consulting, training, and regulatory compliance reporting.

Asset integrity

The rapid increase of internal and external inspection techniques and data analysis technologies has enabled risk-based asset integrity, as discussed in more detail later in this text.

Construction and maintenance

Upgrading and refurbishing mains (both high pressure and distribution) and related equipment, as well as installing and maintaining service lines, are tasks normally performed by operations personnel (fig. 6–14).

Other construction and maintenance functions, such as installing meters and designing and managing large, complex projects, are tasks normally performed by the customer service and engineering groups, respectively. Chapter 9 contains details regarding maintenance and chapter 12 covers construction.

Fig. 6–14. Installing a 2-inch plastic distribution main. This main will replace an existing steel main. First the ditch is dug, exposing the steel main, then the section scheduled for replacement is cut out, and finally the new plastic pipe is tied in.

Inspection

LDCs often employ construction contractors to handle at least some of the construction and maintenance activities. Inspectors work with the contractors and are normally company employees or former construction company personnel with years of experience and specialized training. Their responsibilities include safety and quality assurance. As the work progresses, they check compliance with company specifications, industry standards, design requirements, and regulations. Depending on job size, inspectors may stay with only one large job from start to finish or may oversee several smaller jobs at the same time. Inspectors need knowledge and expertise but also diplomacy as they cajole, wheedle, and threaten, all in the name of safety and quality.

Corrosion control

While it seems dirt should be dirt, there are actually many types of soils. To make matters more difficult, there are many objects on or just below the surface, such as electrical cables, pipes, foundations, basements, and railroad lines, to name a few.

We learn in high school chemistry that solids, liquids, and gases are composed of elements. Each element has its own *electrical potential*, which is essentially its willingness to give up electrons to other elements. Batteries work because of this phenomenon, but it is also responsible for the process of corrosion. Chapter 9 has more about corrosion. For now, it is sufficient to say that small electrical currents, traveling through the ground between objects, cause rust.

The electrical potential of the ground surrounding the pipeline compared to the electrical potential of the pipe itself, called the *pipe-to-soil potential*, varies along the line. Corrosion technicians collect pipe-to-soil readings at test stations (as many as several hundred) installed along the grid.

Corrosion engineers and technicians analyze pipe-to-soil potential readings and employ other diagnostic tools, including close interval surveys (CIS) and alternating current (AC) and direct current (DC) voltage gradients. Based on their analysis and experience, maintenance programs are designed to control corrosion. These programs may involve installing ground beds, rectifiers, or anodes, repairing the pipe coating, or even replacing the pipe. Lack of corrosion is the big advantage of plastic pipe, making all of the above unnecessary.

Emergency response

Any emergency calls for immediate action. The emergency response group develops plans detailing employees' roles and responsibilities when an emergency happens. Similar to the military, they conduct mock drills based on realistic scenarios to test the plans. Lessons learned during the tests are incorporated into the next set of plans. However, the drills do not necessarily involve only company personnel. Often emergency response officials, regulators, support contractors, and even the media are involved for the sake of realism and coordination. Some drills are announced in advance, while others are unannounced, adding an additional element of realism and tension. Emergency response is not governed by time of day (fig. 6–15).

Fig. 6–15. Nighttime leak repair. A homeowner rented the backhoe on the left and proceeded, without calling authorities, to replace a home drain line. Just as he was finishing, he struck a 2-inch plastic distribution main. Local emergency response personnel secured the area just as the leak repair crew was arriving.
Courtesy: Wayne Fairley.

Regulatory compliance

Since its inception, the natural gas distribution industry has faced many safety and economic regulations. In the early days, there were municipal safety regulations, but they focused on the more dangerous gas manufacturing sites, rather than the less-explosive distribution systems. A major step toward national safety regulations occurred when President Lyndon Johnson proposed during his 1967 State of the Union Address that Congress pass nationwide safety legislation. The safety concerns were driven partly by pipeline accidents, including a natural gas transmission line accident in Natchitoches, Louisiana, that resulted in 17 fatalities. The Natural Gas Pipeline Safety Act of 1968 was enacted in response. As regulations grew, so did the need for staff to ensure compliance. Both transmission and distribution pipeline operators have seen a large increase in new regulations in the last 10 years. Most safety regulations use industry-developed codes and standards as their basis. Company employees read the regulations and design programs to achieve and document compliance. These programs and how companies implement them are subject to audit by federal or state officials.

Economic regulations are provided by the Federal Energy Regulatory Commission (FERC) for pipelines that cross state lines, while most distribution pipelines are regulated by state public utility or public service commissions. Some local distribution rates are even regulated at the city level.

Training

Employees of the training department work with managers, supervisors, the most experienced employees, and knowledgeable pipeline consultants to understand and document the knowledge needed to perform each task safely and efficiently. With this knowledge identified, training departments procure existing training, or in some cases, custom-designed training to meet specific needs.

Programs combining online, classroom, and hands-on training are designed to achieve the best training results. Employees receive training and complete testing in order to be qualified to perform the specific task or tasks. Some tasks require periodic refresher courses, while others do not. The training department keeps track of required training and who has attended, sending out schedules and setting up additional opportunities to attend class, work online, or work with experienced trainers as required.

Customer Service

Operations personnel keep the gas moving 24 hours per day, seven days per week, 52 weeks per year. They work with contractors, regulators, local officials, landowners, and transmission company workers, but not normally directly with customers. That is left to customer service.

Call center

The first line of customer service, an LDC call center, looks like most any other call center. People sit at desks and answer phones. Some calls are simply requests for information, such as, "How do I get service connected or disconnected?" At other times people are irate that their service is out and call to ask when it will be back on. Emergency calls also come to the call center, requiring immediate response.

Sorting out and dealing with hundreds or thousands of incoming calls each day, some urgent and others simply information requests, dictates a careful blend of the efficiency of automated response and the effectiveness of personal attention. Most people are satisfied with prompt, accurate, automated responses to information requests. Other calls, like complaints, require careful, personal attention. Most call centers track and maintain effectiveness measures, helping them understand and improve service. Empathetic and knowledgeable call center employees are unsung heroes of customer relations.

Dispatch

Work orders, written or electronic, tell service technicians where to go and what tasks to do. Routine work orders include installing or replacing meters, turning on or off service, and relighting home appliances following a service interruption, to name a few. Emergency work orders like responding to known or suspected leaks, or perhaps homeowners' reports of strange smells, take precedence over routine work orders.

Service

Call centers are the first line of contact with customers, but customers only talk with and never see call center workers. Service workers interact with customers in person in response to odor complaints, to relight appliances following service disruptions, and to replace and maintain meters. Service workers also install new meters and perform a variety of other functions. Depending on the company, the line of demarcation between service workers and other LDC employees is somewhere between the service connection line tap on the service main and the meter. Dispatch and local practices decide who will do what. Of course in an emergency, everyone works together.

Meter reading

Some areas have not yet moved to smart meters and automated meter reading, so some meters are still read manually. Meter readers walk or drive from meter to meter, downloading or manually entering information into special handheld portable computers. Sometimes the readings are transmitted electronically directly from the device. At other times readings are transmitted over phone lines after the devices are plugged into their cradles at service centers. Whatever the method, accurate readings are important as they form the basis of the customer's bill (and the company's revenue).

Customer billing

How customers are charged depends on regulations and practice and varies accordingly. In the United States, most gas bills have two components:

- Cost of gas component for the natural gas used by a customer
- Cost of service component, or the utility's cost to operate the gas distribution system and earn a return on its investment in the system

In some cases customers pay the utility for distributing the gas separately from paying the gas provider.

The gas cost may include a base charge and a "purchased gas adjustment" or a "cost of gas adjustment." The adjustment is essentially a true-up mechanism that adjusts for minor undercollection or overcollection of costs for natural gas purchased on behalf of the utility's customers. This adjustment goes up and down based on the total price paid for the gas purchased in comparison to the actual revenues collected in prior months. Some states require monthly adjustments, others quarterly or semiannually. More frequent adjustments mean gas price changes in the marketplace show up more quickly on customers' bills.

The cost of service component varies by customer class. Different regulators classify customers differently. In general there are four customer classes: residential, commercial, industrial, and electrical power generation, each with its own rate. Industrial and electrical power generators usually pay lower service fees for each unit of gas purchased than residential customers. Serving small customers with uneven demand generally costs more per unit than serving large-volume, steady customers.

Costs are usually split into two parts: a monthly fixed (or meter) charge, and a separate volumetric charge per unit of gas delivered. In total, these charges should generate revenues adequate to cover a utility's operating expenses, plus a reasonable return on their pipeline and other plant investment. Some regulators like to see higher fixed charges and lower volumetric charges. Other regulators, seeking to protect low-income customers, will argue for lower fixed charges and higher volumetric charges. There are solid arguments on both sides of this debate.

The customer billing department does not set the rates. They simply populate computer programs with the predetermined rates, customer classes, purchased gas adjustments, and other special considerations, like low-income assistance. They add to that usage information from meter readers and print out bills.

Collections

There are always some who cannot, and others who will not, pay their bills. Collections work with these people to sort out which is which. They devise payment plans and assistance for some and encourage others. Sometimes, as a last resort, gas service is shut off.

Utilities calculate their average unpaid bills, called *bad debt*, and add this cost to their operating expenses calculation when justifying their rate increase requests to regulators. Utility regulators consider state rules and the utility's collections performance compared to peer companies when deciding what is fair and equitable to both the utility and its customers.

Advertising and marketing

Some people (forgetting about competition from electricity) say regulated monopolies do not need to advertise and in fact should not be allowed to. Electricity took away local distribution's legacy market, street lighting, about a century ago, and now competes vigorously against gas for home heating and cooking. Most LDCs do some limited advertising to attract customers. They also insert mailings into gas bills to get the message out. But some of the best gas sales people are the local operations and service technicians. They know the developers and regularly work with them to get gas distribution mains installed while subdivisions are under construction.

Public relations

Meters, pressure regulators, markers, and other aboveground pipeline components are so familiar most people pay no attention to them. Natural gas service is so reliable few people give it a second thought. Public relations then must educate the public about pipelines in general, and the individual company in particular. Safety is a particular focal point for public relations, so educating the various public audiences (homeowners, excavators, and land use and emergency response officials) is critical. Efforts focus on explaining how to avoid, recognize, and react to emergency situations, such as leaks, and how to prevent outside force damage. To this end, public relations employs standard communication techniques, print, television, public meetings, and increasingly the Internet. In addition to safety, other messages such as corporate citizenship and concern for the community and customers are key messages.

Gas Supply

The tasks seem straightforward: forecast the amount of gas needed, purchase the gas and transportation for it, and ensure it shows up on time, every time, and at the lowest cost. Juggling may seem straightforward also, until a person tries it. Such is the life of the gas supply group, as will become abundantly clear in the next several sections.

Demand forecasting

Forecasting techniques may be very similar, but past experience, personalities involved, market characteristics, and computer models are all different. As a result, so is each company's forecasting technique.

Forecasts are updated yearly for 5, 7, or even 10 years. Emphasis is on the next 12-month cycle, which starts at the end of the winter when the amount of gas in storage is at its lowest level. In the Northern Hemisphere, the 12-month forecast period is normally April through the next March. In the Southern Hemisphere, it may be October through the next September.

Many factors are considered, such as history by customer rate class, residential heating, residential nonheating, industrial, and commercial use, for example. Single large users such as power plants or factories are considered individually and then adjusted for known changes. Future demands are adjusted for population growth as well as factories, shopping centers, or other major buildings anticipated. Other factors, such as unusually warm or cold weather, anticipated oil and gas prices, the economy, and even projected legislation and regulations are all considered in developing the forecast. This first cut at the forecast yields *average demand*, the total amount of gas required for the upcoming year.

However, "average" in the sense of meeting the needs of most of the people most of the time is not good enough. On the coldest day, everyone wants to keep warm. Enter the *design day*. Forecasters study past history to determine the one day of highest total system demand. Not surprisingly, this is normally the coldest day in past history. The design day may have been 20 years in the past, so it is adjusted to current total demand levels. With average and maximum demand in hand, the juggling begins.

Purchasing

The challenge is to purchase enough gas and transportation capacity to ensure gas is always available when needed while leaving enough flexibility to opportunistically purchase gas and transportation if prices decline. Before moving on to the purchasing process, a few words about purchasing in general are appropriate.

LDCs purchase natural gas as a commodity from either producers or marketers, and they purchase transportation as a service from natural gas pipelines. LDCs then provide transportation services from the city gate to the final destination, charging on-system end users a bundled price for transportation and natural gas on a single invoice. Complexity enters the scene as producers or marketers sometimes sell directly to off-systems end users. Depending on economics and logistics, either natural gas transmission pipelines or local distribution pipelines may deliver to the off-system users. Finally, marketers sometimes include a portion of the transportation costs bundled with the price of the gas (fig. 6–16).

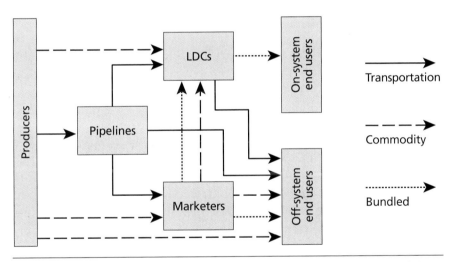

Fig. 6–16. Natural gas purchasing and transportation schematic. Producers sell the commodity and pipelines move the gas. Sometimes purchasers bundle the gas with transportation when they sell it to the next purchaser in line, and the producers do so also.

Purchasing natural gas, transportation, and storage services is an iterative process. It starts with determining gas needs: total, design day, and monthly. Then bid packages are developed and sent to transmission pipelines and storage companies. (Most storage is owned by transmission pipelines, their affiliates, or the LDC itself, but many gas marketing companies lease large blocks of gas storage space and sublease it in turn to their customers.) It seems like the process should be the other way around, gas first and then transportation. However, as one division vice president of gas supply said, "We want an understanding of transportation possibilities before we ask for gas bids, since getting the physical pipeline connections and reasonable transportation prices are more difficult than buying the commodity."

Transportation and storage. With most things, service type and competition determine the rate. Pipeline and storage services are no exception. Over the years, transmission companies, their customers, and regulators have hammered out a list of service categories and rates. More details regarding rates are provided later in this text, but the following list of rates provides a glimpse into their design and complexity:

- Firm transportation service (FTS)
- Flexible firm transportation service (FFTS)
- Interruptible transportation service (ITS)
- Delivered firm storage service (DSS)
- Firm reverse storage service (FRSS)
- Nominated storage service (NSS)
- Best efforts storage service (BESS)
- Interruptible balancing services (IBS)
- Park and loan service (PALS)
- Line pack service (LPS)

The bid packages, sometimes called *requests for proposals*, outline volumes, pressures, delivery locations, services, and other pertinent information. Bids are received. Sometimes they follow the bid outline. Often they contain alternative proposals. Once received, gas supply analysts sift through the proposals, creating spreadsheets or populating analysis databases. With spreadsheets and analyses in hand, gas supply moves to the next step, purchasing the gas.

Natural gas. Gas wells normally produce at fairly steady rates, unless extreme cold weather causes freeze-offs due to moisture in the gas stream, other operating upsets occur, or well maintenance is conducted. However, peak winter gas demand is often twice that of spring or fall. With the risk of winter freeze-offs and higher demand, and obeying the law of supply and demand, gas prices are often higher in the winter than in other seasons. Customers expect gas when they want it and at manageable prices. Gas supply departments build a supply portfolio to guarantee volumes and limit price volatility, while at the same time keeping costs down. Some of the purchasing strategies used to build the portfolio include the following:

- Contract length
- Pricing terms
  - Fixed for a defined term
  - Indexed and adjusted as the index changes monthly or weekly
  - Spot at the current price or next month price
  - Physical hedging through gas futures contracts
  - Financial hedging through gas price options
- Delivery flexibility
  - Guaranteed delivery
  - Interruptible supply
  - Daily call gas

As with transportation and storage, bid packages go out to multiple gas suppliers, who in turn study the packages carefully, hoping to maximize their own returns. Eventually they submit proposals.

Gas supply analysts add these proposals to the stack of transportation and storage proposals and buckle down, working away at the jigsaw. Painstakingly they construct a supply, transportation, storage, and pricing portfolio aimed at satisfying regulators, customers, and management. Following management reviews and approvals (regulators do not approve in advance; they audit after the fact), contracts are signed and the task turns to managing them so that the painstakingly constructed plans become reality.

Scheduling

Weather and history are two of the most critical variables schedulers consider as they determine daily and weekly demand forecasts. They adjust history to reflect anticipated demand changes, such as new subdivisions and malls, and check weather forecasts as they build short-term demand profiles. Computers have replaced paper, pencils, and calculators as schedulers merge demand with supply. Transmission pipelines moving the gas and suppliers putting it into the transmission lines are key considerations. Firm transportation and firm gas are the first factors fed into the supply plan. Spot gas, call gas, and storage are all considered.

Just because the next day's final supply plan is called "final" does not mean it is. In any event, schedulers send their nominations to transmission pipelines and gas suppliers, who build their own schedules accordingly. The whole scheduling and transportation process is tightly integrated and iterative, requiring endless phone calls, e-mails, and adjustments as conditions change. The need for cooperation and coordination spawned formation of standards organizations, which in North America is the North American Energy Standards Board. These organizations bring together industry participants to develop standards and guidelines such as the starting and ending times of the gas day and the timing of nomination cycles. Regulators often adopt the standards, turning them into law. Finally, schedules are sent to gas control for implementation. Figure 6–17 is an example of what gas control receives from gas supply.

Meters	BTU Factor	Shipper 1	Shipper 2	Shipper 3	Shipper 4	Total MMBTU	Total MCF
Pipeline 1							
Delivery Point 1	1.0140				—	—	—
Delivery Point 2	1.0045				7,500	7,500	7,466
Delivery Point 3	1.1000				30	30	27
Total					7,530	7,530	7,494
Pipeline 2							
Delivery Point 1	1.0050				875	875	871
Delivery Point 2	1.0200				626	626	614
Delivery Point 3	1.0200				—	—	—
Delivery Point 4	1.0200		1,250			1,250	1,225
Total						2,751	2,710
Pipeline 3	1.0250				500	500	488
Pipeline 4	1.0262				1,000	1,000	974
Pipeline 5							
Delivery Point 1	1.0250		1,345		25	1,370	1,337
Delivery Point 2	1.0250	225		350		575	561
Delivery Point 3	1.0250				1,245	1,245	1,215
Delivery Point 4	1.0250		25			25	24
Delivery Point 5	1.0250				625	625	610
Delivery Point 6	1.0250				445	445	434
Delivery Point 7	1.0250				1,200	1,200	1,171
Delivery Point 8	1.0250				—	—	—
Total		225	1,370	350	3,540	5,485	5,351
Grand Total		225	1,995	350	12,571	16,516	16,286

Fig. 6–17. Flow requirements and expected flow by pipe. The schedule lists expected receipts for the day from each pipeline at each city gate. In this case Shipper 4 is the LDC. The other shippers move on the LDC system to various customers. Volumes (MCF) are measured and total British thermal units are calculated.

Contract management

A detail-oriented, dedicated band of accountants and analysts compare multiple receipt and delivery tickets from each city gate to contracted natural gas and transportation volumes. Daily, monthly, and sometimes even hourly, adjustments are commonplace. If there is not enough volume, the company may have to pay anyway. Moving more volumes than forecast may result in excess charges. Computers, databases, and various analytic tools help, flagging discrepancies, but in the end it all comes down to human judgment.

To further complicate matters, regulators routinely ask for audits, adding additional work. One LDC estimates approximately 20% of their records are audited each year by the regulator.

Other Groups

As shown previously in figure 6–12, two other groups, finance and administration, play vital roles. They handle tax, treasury, accounting, insurance, legal, human relations, purchasing, and a myriad of other duties quite similar to how those functions are handled in most large companies. Two special functions—rates, filings, and compliance, and right-of-way—warrant special mention.

Rates, filings, and compliance

LDCs only make money on the local transportation service they provide, not on the gas they carry. They pass the cost of the gas along to their customers, but the total cost must be the same cost the LDC pays. They cannot add any markups or extra charges. Regulators and consumer watchdog groups carefully monitor the company's records. The company purchases gas from a variety of sources at different prices and must allocate these prices to customers by location.

Gas pricing. Transmission line constraints to one service area, for example, might make gas supplies tighter and prices higher. Another area supplied by the same LDC might have plentiful supplies and lower prices. Regulators decide whether the gas costs for each area will be kept separate and charged accordingly, or instead will be added together and averaged. Either method gives the LDC the same revenue, but the costs will vary for the customers in the two service areas.

The rates, filings, and compliance group, or a similarly named group, works closely with both the gas supply and accounting groups to keep track of how much gas is bought, its price, and where it goes. Costs are allocated accordingly. Large service areas, all paying the average price of gas supplied to the area, make life simpler for these groups than small service areas, each with different gas costs.

Transportation rates. Customers pay LDCs to transport natural gas from city gates to their homes. Most, but not all, LDCs are essentially monopolies. They operate under a charter or franchise from the city or state. The exclusive right to serve an area comes with a price tag— economic regulation and rate caps. Most regulators employ a cost-of-service concept when determining allowable maximum rates. This approach allows LDCs to recover their costs and in addition earn a return on their investment.

Rates, filings, and compliance is normally a small group of financial and legal specialists. They develop documents asking for and justifying rate increases and file these documents with the proper agency. Of course the rates are seldom accepted as filed, so the group, with legal and management assistance, negotiates with regulators and usually other interested parties until they either reach agreement or go to court.

Right-of-way

Most pipelines are buried. Owning the surface under which they are buried is not necessary. Having the right to be buried there and the right to keep others from building permanent structure above the pipeline is. This right, commonly referred to as *right-of-way* or *ROW*, is negotiated and managed by the right-of-way department or group. Sometimes the ROW is purchased from private landowners; more often it is granted to the LDC by the city or developer. Most LDCs have more than 90% of their pipelines in easements provided by the cities they serve. However obtained, the right is documented in agreements that are then filed with the city or county government. The right-of-way group manages ROW acquisitions and handles matters associated with land rights as they arise.

Summary

- Local distribution lines are composed primarily of city gates, high-pressure mains, distribution mains, and service lines.
- Pressure-reducing stations along the way decrease pressures to the levels needed to supply the customer, normally less than 60 psi.

- More than 90% of new lines are plastic.
- Supplemental gas for peak shaving comes from storage and from LNG and LPG plants.
- Systems must be designed with enough capacity to handle peak load on the coldest day of the year.
- Operations keeps the gas flowing; maintenance keeps the system working.
- Natural gas demand fluctuates widely during the day and between seasons.
- Leak surveys are conducted regularly to find and repair leaks.
- Gas only ignites in air at concentrations between about 5% and 15%.
- Operating and repair manuals provide training and guidance.
- Plastic pipe eliminates corrosion concerns.
- Call centers are the primary customer connection point.
- Gas supply forecasts needs and uses a variety of purchasing methods to keep prices down, while at the same time ensuring reliable supply.
- Transportation services and natural gas are normally purchased separately.
- Scheduling is a never-ending, iterative process.

Notes

1. PHMSA Form F 7100.1-1 (2006), https://hip.phmsa.dot.gov/analyticsSOAP/saw.dll?Portalpages (requires login).
2. American Gas Association, "System Type: Distribution," *Knowledge Center, Natural Gas 101*, https://www.aga.org/knowledgecenter/natural-gas-101/natural-gas-glossary/s.
3. Gas Technology Institute, "Plastic Pipe Services: The Source for Infrastructure Safety and Reliability," *Gas Operations News*, February 2006, vol. 3, no. 1, http://www.gastechnology.org/news/Documents/Gas_Ops_News/GasOpsNews_Feb2006_Fnl.pdf.
4. American Gas Association, "Peak Shaving," *Knowledge Center, Natural Gas 101*, https://www.aga.org/knowledgecenter/natural-gas-101/natural-gas-glossary/p.

chapter 7

Liquefied Natural Gas

The person who knows "how" will always have a job.
The person who knows "why" will always be his boss.

—Diane Ravitch

    The chief LNG operator, with years of plant experience, sits in the plant control room with a newly minted process engineer. The engineer has a head full of thermodynamic equations and is eager to learn. As the plant hums along, converting the gaseous natural gas into liquefied natural gas, they discuss plant operations.

    The chief operator explains the plant uses one of several proprietary natural gas liquefaction processes. Regardless of the process, prior to liquefaction, dust, pipeline debris, water vapor, carbon dioxide, hydrogen sulfide, mercury, and other contaminants must be removed from the gas stream. Otherwise they might freeze and plug up or otherwise interrupt the process. After contaminants are cleaned from the stream, residual ethane, propane, butane, and other heavier (larger) hydrocarbon molecules are removed from the stream as it is compressed, cooled, and successively run through the deethanizer, debutanizer, and depropanizer.

    Finally, the remaining methane (with perhaps a little ethane) enters huge refrigeration systems that chill the stream to about −160°F. The engineer nods his head with understanding until the chief operator says, "And this is where the magic happens. At this valve the pressure is reduced, turning the gas into a liquid." This statement defies the engineer's book knowledge of thermodynamics. The chief operator claimed reducing the pressure turned the gas into liquid. But the engineer "knows" high pressure, not low pressure, liquefies gas. The engineer's understanding of thermodynamics and the operator's knowledge of the plant clashed.

Who is right? Well, to some extent, both are. Whether natural gas is liquid or gas depends on both temperature and pressure. The step described by the chief operator is used to cool natural gas, but no magic is involved. Instead, one of the laws of physics, the conservation of energy, is at work across a Joules-Thompson valve. The chief operator does not know the stream was already liquid well in advance of entering the valve.

During the pressure drop across the valve, some of the LNG turned back into gas (flashed or boiled off), making the remaining liquid even colder. The gas is flashed several times during the process until it is cold enough to store as LNG in an insulated tank at a relatively low pressure.

> **Joules-Thompson Valve**
>
> A device designed to convert liquid to gas across a small distance by rapid pressure reduction. This conversion requires energy. Since no energy is added to the process, the temperature drops.

What Is LNG?

So what is LNG, anyway? While LNG could stand for liquid natural gas, normally LNG stands for liquefied natural gas. Simply put, LNG is the same substance used to heat many homes, office, schools, hospitals, and places of business. It is composed of mainly methane, but in liquid, rather than gaseous, form. Natural gas transmission and distribution systems are designed to operate with gaseous methane at ambient temperatures, so LNG must be regasified prior to entering the natural gas grid.

A Brief History of LNG

In 1873, Karl Von Linde built a compression refrigeration machine in Munich. Construction of the first LNG plant began in the United States in West Virginia in 1912, and operations commenced in 1917. The first commercial liquefaction plant was built in Cleveland, Ohio, in 1941. The first marine transportation of LNG occurred in 1959 when the *Methane Pioneer* transported a cargo from Lake Charles, Louisiana. Following this demonstration of marine transport, the British Gas Council began importing LNG into the United Kingdom. Discovery of the North Sea gas fields meant importation of LNG was no longer needed. Other countries, however, began to construct import and export facilities.

Why Liquefy Natural Gas?

Why liquefy natural gas only to later regasify it prior to use? The reason is that it requires less space. Natural gas demand is cyclical, between seasons and even within a given day, so it must be stored between seasons and from hour to hour within the day. Natural gas is sometimes produced long distances from the point of consumption. When those distances are across land, gas is moved by pipeline. Liquefying gas makes marine transportation of LNG considerably more efficient, as the liquid requires less space during transportation.

To amplify the point, transporting 1 gallon of water from home to office as a liquid in a 1-gallon container is easier than transporting the same gallon as steam in a 600-gallon container. In the same way, 600 cubic feet of natural gas at standard temperature and pressure can be stored as 1 cubic foot of liquid. Natural gas liquefaction makes marine import and export facilities economic. Figure 7–1 shows a typical LNG export terminal.

Fig. 7–1. LNG export terminal. The liquefaction plant is in the foreground. Farther back are the storage tank, ship loading, and LNG vessel.
Courtesy: ConocoPhillips.

The Science behind LNG

Whether a substance is in liquid, solid, or gaseous form depends on the temperature and pressure and varies by substance. Water, for example, is liquid between 32°F and 212°F (0°C to 100°C) at sea level. Move to a higher elevation, however, and water boils at less than 212°F.

Conceptually then, methane can be cooled and compressed to the point where it liquefies. However, methane, like other substances, has a critical pressure and critical temperature. Above the critical temperature of methane, –117°F (–83°C), it will not liquefy regardless of pressure applied. So, it must be cooled below its critical temperature before it will liquefy. The critical pressure of methane, the pressure at which it liquefies at critical temperature, is 670 psi (4.6 MPa). In other words, at –117°F and 670 psi, methane turns from gas to liquid. When the temperature is reduced below –117°F, less pressure is required to keep the methane in liquid form. LNG is normally cooled to between –250°F and –260°F, where the pressure required to contain it is quite low.

In summary, the science of LNG is quite simple. No chemical changes happen. The gas is simply cooled to the point it becomes liquid. The lower the temperature, the lower the pressure required to keep the methane liquid.

Gas → Liquid → Gas

Taking methane from gas to liquid and back to gas starts as the gas stream is received from transmission pipelines or the production field. It requires five steps before the gas stream is put back into a transmission line or is released directly to distribution systems (fig. 7–2).

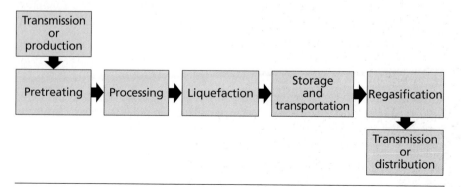

Fig. 7–2. From gas to liquid to gas

LNG plants consist of one or more *trains*. Each train has its processing equipment connected in series. Small capacity expansions are accomplished by optimizing individual units in the train. Major expansions require installing another complete train.

Pretreating

The amount and types of pretreating required depend on gas quality. LNG plants using natural gas previously treated and moved by transmission lines may only require limited pretreating to remove dust, pipeline debris, and a little water vapor. Natural gas streams entering export facilities may require extensive pretreating, including removing any condensed NGLs, CO_2, H_2S, water, mercury, and perhaps other contaminants that would impact plant performance.

Processing

Although condensed natural gas liquids are removed during pretreating, some ethane, butane, propane, and heavier hydrocarbons may remain in the stream in gaseous form, especially at LNG export terminals. These heavier hydrocarbons often have a higher value than methane, if a market for them exists. When there is no ready market for the heavier hydrocarbons at the export location, those supplying the gas stream to the export facility sometimes attempt to leave as much of the heavier hydrocarbons in the stream as possible to avoid the cost of extracting them. The varying energy content driven by the amount of heavier hydrocarbons left in the stream presents two problems.

Gas interchangeability. Water heaters, stoves, ovens, furnaces, power generation stations, and other users of natural gas are designed to operate within a given energy range. When the Btu content goes above or below the range, various problems occur, such as increases in soot, carbon, and pollutant emissions. Furthermore, a shortened heat exchanger life could result, along with extinguished pilots, reduced performance, and damage to heat transfer equipment. Thus gas interchangeability becomes an issue.

Interchangeability is defined as "the ability to substitute one gaseous fuel for another in a combustion application without materially changing operational safety, efficiency, performance or materially increasing air pollutant emissions."[1] In the early 2000s, the United States became concerned with the energy content of what were at that time projected to be significant LNG imports. This issue is less pressing now due to the rapid growth of natural gas production from unconventional sources, including shale formations.

Gas interchangeability can be measured with various indexes, the most common of which is the Wobbe index, sometimes referred to as the *interchangeability factor*. The definition of the Wobbe number is based on the heating value and specific gravity of a gas, and it is related to the thermal input to a burner (Btu per hour).[2] While attention to this topic in the United States has subsided, it remains an important consideration.

LDC peak-shaving liquefaction operations. Natural gas demand varies by season and throughout the day. Seasonal demand is met primarily by underground storage, which is often located far from the point of consumption. Interday demand swings are most conveniently handled by storage close to the consumption site. To serve these interday demand swings, some LDCs use LNG plants designed for the natural gas stream anticipated (fig. 7–3).

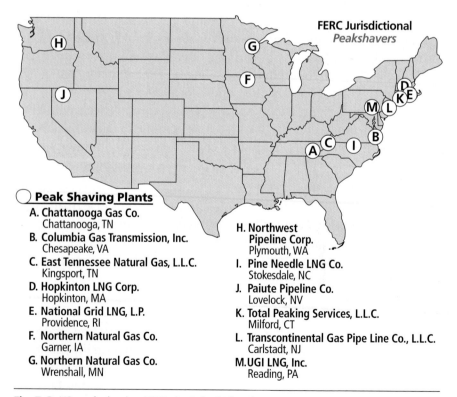

Peak Shaving Plants
A. Chattanooga Gas Co.
   Chattanooga, TN
B. Columbia Gas Transmission, Inc.
   Chesapeake, VA
C. East Tennessee Natural Gas, L.L.C.
   Kingsport, TN
D. Hopkinton LNG Corp.
   Hopkinton, MA
E. National Grid LNG, L.P.
   Providence, RI
F. Northern Natural Gas Co.
   Garner, IA
G. Northern Natural Gas Co.
   Wrenshall, MN
H. Northwest
   Pipeline Corp.
   Plymouth, WA
I. Pine Needle LNG Co.
   Stokesdale, NC
J. Paiute Pipeline Co.
   Lovelock, NV
K. Total Peaking Services, L.L.C.
   Milford, CT
L. Transcontinental Gas Pipe Line Co., L.L.C.
   Carlstadt, NJ
M. UGI LNG, Inc.
   Reading, PA

Fig. 7–3. US peak-shaving LNG plants built for short-term storage
Source: Data from US FERC, "LNG," http://ferc.gov/industries/gas/indus-act/lng/peakshavers.pdf.

When energy content changes because heavier hydrocarbons are in the stream, these existing plants may require extensive retrofits.

Liquefaction

Several different liquefaction processes exist, including the following:
- Pure refrigerated cascade
- Propane-precooled mixed-refrigerant
- Precooled mixed-refrigerant, with back-end nitrogen expander
- Other mixed-refrigerant
- Nitrogen expander-based[3]

The processes use different refrigerants and methods to cool the gas, but the goal remains to cool it to the point where it condenses into a liquid. Figure 7–4 shows a typical LNG plant.

Fig. 7–4. Typical LNG plant processing equipment
Courtesy: ConocoPhillips.

Storage and transportation

Storing LNG involves a delicate (but straightforward) balance of pressure and temperature. The colder the methane, the lower the pressure needed to keep it in liquid form. Lower temperatures mean lower pressure, and lower pressure means a thinner (less-expensive) pressure vessel.

LNG storage tanks are heavily insulated to keep the LNG as cold as possible. They are never filled completely with LNG. Some vapor space is always left above the liquid. When ambient heating causes the temperature of the LNG in storage to rise, the vapor space above the LNG, which is filled with natural gas, obeys Charles's law. Accordingly, pressure rises with the temperature. In response to the rising pressure, LNG station controls remove natural gas vapors from the tank, thereby reducing the pressure in the tank. When the tank pressure falls below the vapor pressure, some of the LNG boils off into a gas, absorbing energy, cooling the liquid. Boiling off the LNG to gas maintains optimum internal tank temperature and pressure. The boiled-off gas is typically used for fuel at the plant or on the ship. Alternatively, it may be recooled, reliquefied, and returned to the tank.

Tanks. LNG tanks take their cue from thermos bottles. They have two concentric walls with insulation between to minimize the effects of ambient heat. The inner tank is usually constructed of 9% nickel steel, and the outer tank is usually constructed of carbon steel or concrete (fig. 7–5).

Fig. 7–5. LNG storage tank. These two 144,000 cubic meter tanks were built more than 30 years ago but were never used due to public concerns over LNG.
Courtesy: Steve Vitale.

The tank floors and roofs are also insulated. It is interesting to note the tank floors normally have heaters underneath them to protect the earth under the tank from freezing, thus causing frost heave problems. Various types of containment systems are used to capture LNG in the event of an inner wall failure.

Ships. Ocean-going vessels are specifically designed and constructed as dedicated LNG carriers. Ship designers use one of two tank designs. Most people, when they envision LNG tankers, think of a ship with a number of domes lined up longitudinally on the deck. These are actually spherical tanks developed by Moss Maritime and thus are called *Moss tanks* (fig. 7–6).

Fig. 7–6. LNG tanker with Moss-type tanks
Courtesy: Moss Maritime.

As with onshore storage tanks, vessel storage tanks are constructed with two walls with insulation.

Membrane tanks consist of two containers, one inside the other, built of thin layers (membranes) of engineered metal. The membranes have insulation between them and sometimes between the outer membrane and the ship's hull.

In both cases temperature and pressure are controlled by boiling off, as with the shore tanks. LNG vessels are often *dual fueled*, meaning they can use the boiled-off gas as engine fuel.

Regasification

Prior to entering the natural gas transmission system in the case of import terminals, or the distribution system in the case of peak-shaving plants, LNG is converted back into natural gas. Again, pressure and temperature come into play. LNG is removed from storage and directed through a series of heat exchangers. As with liquefaction, there are several different processes used to regasify or vaporize LNG. These processes include the following:

- Open rack (ORV)
- Submerged combustion (SCV)
- Shell and tube (STV)
- Ambient air (AAV)
- Combined heat and power (CHP)[4]

Onshore regasification plants have been in use for many years, but recently there has been a growing demand for floating storage regasification units (FSRUs). FSRUs are often retrofitted LNG tankers. LNG is transferred to the FSRU, where it is vaporized and sent ashore.

Export and Import Terminals

As of 2012, according to the Bureau of Economic Geology at the University of Texas, there were 25 LNG export terminals, 91 import terminals, and 360 LNG ships, together handling approximately 220 million metric tons of LNG every year.[5] At the export terminal, LNG is processed, liquefied, and loaded onto ships. Import terminals offload, store, regasify, and inject natural gas into transmission lines (fig. 7–7).

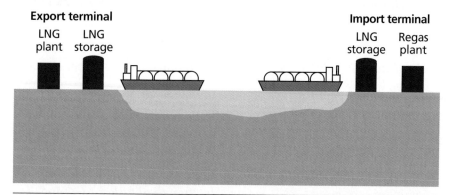

Fig. 7–7. Schematic of the LNG value chain

Other LNG Facilities

While LNG export and import terminals receive most of the attention and are the largest in size and volumes handled, it would be a mistake to think they comprise the majority of the facilities. As of 2012, according to the Bureau of Economic Geology at the University of Texas, there were currently about 260 peak-shaving and LNG storage facilities worldwide, some operating since the mid-1960s.[6]

According to the US FERC, there are more than 110 LNG facilities operating in the United States, performing a variety of services. Some facilities export natural gas from the United States, some provide natural gas supply to the interstate pipeline system or local distribution companies, and others are used to store natural gas for periods of peak demand. There are also facilities that produce LNG for vehicle fuel or for industrial use.[7]

Summary

- LNG is the same substance used to heat many homes, offices, schools, hospitals, and places of business. It is composed of mainly methane, but in liquid form.
- LNG must be regasified prior to entering the natural gas grid.
- Natural gas at standard temperature and pressure occupies about 600 times more volume than LNG.
- Taking methane from gas to liquid and back to gas requires five steps: pretreating, processing, liquefaction, storage, and regasification.
- The amount and types of pretreating required depend on gas quality.
- Gas interchangeability is an emerging issue.
- Proprietary liquefaction processes use different refrigerants and methods to cool the gas, but the goal of each is the same: to cool the gas stream to the point where it condenses into a liquid.
- Storing LNG involves delicate (but straightforward) balancing of pressure and temperature.
- LNG storage containers have two concentric walls with insulation between to minimize the effects of ambient heat.

- Ocean-going vessels are designed and constructed as dedicated LNG carriers.
- Several different processes are used to regasify or vaporize LNG.
- LNG facilities are growing at a rapid rate.

Notes

1. NGC+ Interchangeability Work Group, "White Paper on Natural Gas Interchangeability and Non-Combustion End Use" (February 28, 2005), 3.
2. Ibid., 8.
3. Michael D. Tusiani and Gordon Shearer, *LNG: A Nontechnical Guide* (Tulsa, OK: PennWell, 2007), 112.
4. Ibid., 176–179.
5. Michelle Michot Foss, *Introduction to LNG: An Overview on Liquefied Natural Gas (LNG), Its Properties, Organization of the LNG Industry and Safety Considerations* (Houston: Center for Energy Economics, University of Texas at Austin, January 2007), 3.
6. Ibid.
7. US Federal Energy Regulatory Commission, "LNG," http://www.ferc.gov/industries/gas/indus-act/lng.asp.

chapter **8**

Releases, Leaks, and Leak Management

"This above all: to thine own self be true,
And it must follow, as the night the day,
Thou canst not then be false to any man."

—Polonius in William Shakespeare's Hamlet

Introduction

One of the worst natural gas disasters on record happened on March 18, 1937, at the beautiful, steel-framed, K–12 school building in New London, Texas. The disaster occurred before odorant was routinely injected into natural gas. According to one report, "At 3:17 PM, Lemmie R. Butler, instructor of manual training, turned on a sanding machine in an area which, unknown to him, was filled with a mixture of natural gas and air. The switch ignited the mixture and carried the flame into a nearly closed space beneath the building." The resulting explosion killed 293 people, most of them children. The force was so powerful that "it hurled a two-ton concrete slab 200 feet away, where it crushed a 1936 Chevrolet" (fig. 8–1).[1]

Fig. 8–1. Two-ton concrete slab on a 1936 Chevrolet
Courtesy: New London Museum.

This accident led to regulations requiring the addition of odorants to natural gas, which is normally odorless.

An interesting side note is that one of the first reporters on the scene was a young UP reporter named Walter Cronkite. Later in his career, Cronkite said, "I did nothing in my studies nor in my life to prepare me for a story of the magnitude of that New London tragedy, nor has any story since that awful day equaled it."[2]

Everyone intuitively understands natural gas leaks can be disastrous. This concern is supported by a review of the US National Transportation Safety Board's Web site listing of investigations of pipeline accidents.[3] According to statistics published by the US Pipeline and Hazardous Materials Administration, part of the US Department of Transportation, during the years 2002 through 2011, local distribution lines suffered an average of 32.5 incidents classified as "serious" per year, causing an average of 11.8 fatalities per year over the same 10-year time frame.[4] While an average of 11.8 deaths is 11.8 too many, that number pales in comparison to the thousands of fatalities each year due to traffic accidents. As another frame of reference, *Challenger* and *Columbia* each carried a crew of seven.

Most people find it hard to believe any size leak is "nonhazardous." However, natural gas is only ignitable between about a 5% and 15% concentration in air. Thus small leaks located outside, which are expected to remain the same size and never reach the 5% concentration level, are classified as nonhazardous. The same size leak contained inside a building when the concentration grows to 5%, however, can be deadly. Like the New London disaster, most fatalities associated with local distribution lines occur when gas collects inside a building and ignites. Accordingly, a vigilant public plays a major role in detecting natural gas leaks before the concentration of gas builds to the point it is ignitable.

The vast majority of the more than 25 trillion cubic feet of natural gas consumed in the United States each year travels safely and without incident. The United States enjoys safe natural gas transmission and distribution systems, while meeting consumer demand for relatively cheap energy. When juxtaposed against the spectacular nature and news coverage of natural gas accidents, however, the situation presents challenges to local distribution companies, their industry associations, their regulators, and the public. Together all of these stakeholders seek to prevent leaks and find and repair them before accidents occur, while still keeping consumer costs low.

Releases

Leaks are a subset of a category of events called *releases*. Normal operations require the intentional release of natural gas. For example, some natural gas actuators expel natural gas each time they function. First, gas from the line flows into one end of a cylinder or one side of a diaphragm. Next, the gas from the other end or side is displaced, allowing the piston in the cylinder or the diaphragm contained in the housing to move. This causes the valve to open or close. Some maintenance tasks require evacuating gas from the line to prevent explosions or fires during repair activities. When large quantities must be purged, they can sometimes be directed back into the line. At other times, gas is directed through a flare where it is ignited and burned (fig. 8–2).

Fig. 8–2. Flaring natural gas from a manifold prior to maintenance activities

Smaller quantities may be released to the atmosphere without combustion as the natural course of maintenance activities. In the event of abnormal operations, gas is often released to prevent overpressure situations. Leaks are unintentional releases.

Leaks

Leaks come in a variety of sizes ranging from complete (or nearly complete) separation of the line, commonly referred to as a *rupture*, all the way down to very small leaks, commonly classified as *nonhazardous leaks* (fig. 8–3).

The constituent parts of natural gas are lighter than air. When natural gas leaks, the released gas forms a cloud. If the leak occurs above ground with no wind, the cloud expands from the leak, rising and dissipating into the air (fig. 8–4).

Chapter 8 Releases, Leaks, and Leak Management   209

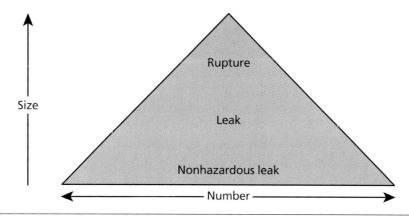

Fig. 8–3. Leak pyramid. The leak pyramid provides a graphical representation of the fact that leak frequency decreases as leak size increases. This same pyramid applies generally to other safety incidents as well.

Fig. 8–4. Gas cloud. The exact pattern of the cloud depends on leak size, geometry, pressure, and various other factors.

As the leak cloud fans out, the amount of methane concentration in the air decreases. The reduction in concentration is important as methane only ignites in a somewhat narrow range of concentrations. The lower flammable limit (LFL), also referred to as *lower explosive limit* (*LEL*) of methane is about 5% concentration in air. The upper flammable limit (UFL), also referred to as *upper explosive limit* (*UEL*) of methane is about 15%. The vapor cloud just a short distance away from small leaks may have a LFL below 5%, which means holding a match there theoretically would not ignite the stream, but *do not* try this.

Wind dissipates gas clouds, and the faster the wind, the quicker the cloud spreads out. Of course, the gas moves in the same direction as the wind. Fluids flow along the path of least resistance, and natural gas is no exception. This means an underground leak may *vent* (come to the surface) directly above the leak, or it may flow some distance before it surfaces. The tendency to migrate makes it imperative to check all available openings, including manholes, sewers, vaults, and the like for indications of natural gas.

One example of an outside leak migrating into a residence is the explosion that occurred on March 5, 2008 in Plum Borough, Pennsylvania. The explosion occurred when natural gas leaking from a crack in a 2-inch steel natural gas line located near the street migrated approximately 35 feet along a sewage line and entered the house. The explosion killed a man, seriously injured a four-year-old girl, destroyed three houses, and damaged eleven other houses. According to the findings, the crack was the result of work performed five years earlier when a master plumber and a hired excavator had replaced the sewage line along which the gas later migrated.[5]

How gas migrates from the leak site to the vent site depends on a number of factors, including the following:
- Soil type and moisture content
- Surface cover, such as yards, streets, and parking lots
- Line pressure
- Depth of burial
- Leak size and age
- Land topographical features, such as slopes and elevation changes

The migratory nature of natural gas means finding the exact leak location, even after the leak is detected, can be challenging.

Detecting Leaks

Natural gas companies conduct regular leak surveys using sophisticated instruments to detect system leaks. Once detected, the exact location of the leak must be pinpointed. In addition to instruments, leaks are discovered by a variety of means. PHMSA publishes the *Guidance Manual for Operators of Small Natural Gas Systems* and addresses leak detection methods.

Leak detection methods

As discussed previously, since the LEL for natural gas is about 5% concentration in air, odorant is added to the gas. The average person can smell the odorant in the gas in concentrations of about 1%, giving a safety factor of 4 times. Although smelling odor means the concentration is at least 1%, it is important to note that it may be significantly higher.

The following are leak detection methods according to the PHMSA:

1. **Odor.** *Gas is intentionally odorized so that the average person can perceive it at a concentration well below the explosive range. . . . The odor of gas may be filtered out as the odorized gas passes through certain types of soil. It may also be modified by passing through soil and into a sewer system containing vapors or fumes from other combustibles as well as the sewer odor itself. Therefore, odor is not always totally reliable as an indicator of the presence or absence of gas leaks. . . .*

2. **Vegetation.** *Vegetation in an area of gas leakage may improve or deteriorate, depending on the soil, the type of vegetation, the environment, the climate, and the volume and duration of the leak. Changes in vegetation may indicate slow below ground leaks.*

3. **Insects (flies, roaches, spiders).** *Insects migrate to points or areas of leakage due to microbial breakdown of some components of gas. Some insects like the smell of the gas odorant. Heavy insect activity, particularly near the riser, the gas meter, or the regulator, can sometimes indicate a gas leak.*

4. **Fungus-like Growth.** *Such growth in valve boxes, manholes, etc., may indicate gas leakage. These fungi grow best where there is a lack of oxygen, such as in a gas leak area. The color of the growth is generally white or grayish-white, similar to a coating of frost. . . .*

5. **Sound.** *Listen for leaks. A hissing sound at a bad connection, a fractured pipe, or a corrosion pit is the usual indication of a gas leak. . . .*
6. **Unaccounted for Gas.** *A possible leak is indicated when an off-peak reading of a master meter with a known average seasonal utilization rate, shows unaccountably high usages rate. Periodic off-peak checks (preferably in the summer months from midnight to 3 or 4 a.m.) can be averaged to provide data for comparison in future checks. This method may indicate a leak on the system, but will not provide a location for the leak. . . .*
7. **Soap Solutions.** *A soap solution can pinpoint the location of a leak on an exposed pipe, on the riser, or on the meter. The solution is brushed or sprayed on and the location of bubbling indicates leakage.*[6]

Soap solutions are often used to confirm a leak when it is initially detected by one of the other methods. Escaping gas will cause a soap solution applied directly on the pipe or fitting to bubble, indicating the presence of a leak (fig. 8–5).

Fig. 8–5. Soap solution bubbling at a corrosion leak in a riser

Leak detection instruments

When using instruments to detect, size, and locate leaks, there are three critical factors:
- Venting (discussed previously in this chapter)
- Instrument
- Operator

There are a number of leak detection instruments, which are detailed below.

Flame ionization detector (FID). Used by the natural gas industry since the late 1950s, FIDs internally combust hydrogen gas, producing a small flame. A pump inside the unit continuously pulls outside air across the hydrogen flame. Any hydrocarbon vapors in the air are burned by the flame. The instrument analyzes the combustion products and indicates the concentration of total hydrocarbons on the meter in terms of parts per million. It serves as a valuable indicator of ignition potential. FIDs are either portable or equipment mounted. Figure 8–6 shows a portable FID.

Fig. 8–6. Portable FID. The scale on the right only shows total hydrocarbon concentration and does not differentiate between methane and other hydrocarbons.

Technicians responding to calls from members of the public who suspect they "smell gas" often use FIDs. However, FIDs cannot differentiate methane from other hydrocarbons. Consequently, they are useful to rule out a leak if no hydrocarbons are present. The presence of hydrocarbons, however, does not necessarily mean methane is present; that calls for a more sophisticated instrument.

Optical methane detector (OMD). Commonly mounted to the front of vehicle for conducting leak surveys, the OMD consists of an infrared light source, receiver, and control box. The infrared light beam shines across the front of the vehicle to the receiver. If the light beam passes through methane, some of the light is absorbed. The instrument control box compares the amount of light received at the receiver with that sent from the light source. It then converts the amount of light absorbed across the path to concentration in parts per million. OMDs allow faster driving and survey speeds than FIDs and are selective to specific hydrocarbons, meaning fewer false alarms. Figure 8–7 shows an OMD mounted to a vehicle conducting a leak survey.

Fig. 8–7. Front-end mounted OMD used to conduct leak surveys
Courtesy: Heath Consultants Incorporated.

Handheld, portable, infrared optical gas detection systems use the same basic technology as the OMD. The main difference is handheld units sample the atmosphere passing through the device, whereas OMDs sample the atmosphere as they pass over the ground. The handheld infrared systems have definite advantages over FID technology as they can discriminate between methane and other gases. Infrared units also do not require gas cylinders and refill systems. Consequently handheld infrared units, which actually look a lot like handheld FIDS, are replacing the longstanding FID instruments as the industry workhorse for home leak detection (fig. 8–8).

Fig. 8–8. Technician conducting leak survey with a portable infrared unit
Courtesy: Heath Consultants Incorporated.

Remote methane leak detector (RMLD). Tunable diode laser absorption spectroscopy is the technology basis for the RMLD. The laser emitter/receiver unit generates a laser beam that is passed over the pipeline. The beam strikes background objects and bounces back to the emitter/receiver unit. Any methane passing through the beam absorbs a portion of the laser light. The control box connected to the emitter/receiver is tuned (calibrated) to only consider the light spectrum absorbed by methane. So, like the OMD, the RMLD can focus solely on methane.

The FID, OMD, and combustible gas indicator (CGI) instruments have to be within, or at least very near, the gas plume before they can detect a leak. The RMLD does not. The laser beam can be pointed at the suspected leak site from as far away as 100 feet and still detect the leak. This feature leads to increased productivity and safety. Leak survey crews appreciate the ability to survey for leaks across a fence when the yard contains a threatening dog (fig. 8–9).

Since the laser beam is light, it will pass through glass, so the RMLD can often be used from outside a building to detect methane inside the building by shining it through the window.

Fig. 8–9. Laser device used to survey for leaks over a fence
Courtesy: Heath Consultants Incorporated.

Combustible gas indicator (CGI). Wires (filaments) coated with a catalyst form the heart of the CGI. As hydrocarbons pass through the meter, a chemical reaction is triggered by the catalyst on the filament. When the chemical reaction occurs, it heats the wire, and the hotter the wire, the higher the concentration of hydrocarbon. Whether or not the chemical reaction occurs depends on the type of catalyst used to coat the filament and the type of hydrocarbon tested.

Thus in the case of CGIs designed to detect and measure methane, the filament is coated with a methane-specific catalyst. Other hydrocarbons do not react as they pass through the instrument, meaning a CGI is used to determine the presence and concentration of methane. Most CGI instruments have a gauge that reports both percentage of flammable gas in the air (percent gas scale) and percentage of the lower explosive limit (LEL) scale. Figure 8-10 shows a technician inserting the CGI probe into a bar hole as he attempts to locate a leak.

Fig. 8–10. Technician using a CGI to read percent LEL as he attempts to pinpoint the location of a leak. The sensor line from the CGI is inserted into a bar hole.
Courtesy: Heath Consultants Incorporated.

In summary, technicians use FID, OMD, and RMLD instruments to conduct leak surveys. Once a leak is detected, they use a CGI instrument to locate its exact position.

Operator

The old saying, "The machine is only as good as its operator," is true for leak detection instruments as well as for any other piece of equipment. According to the PHMSA, "Leak survey technicians must be trained and qualified in the operation of the FI and CGI" or other instruments they are using. "Additional training is required on leak survey procedures, leak classification, recognition of hazards, and pinpointing. All gas personnel should also receive training on 'make safe' actions," or how to secure the area and ensure safety until the leak is repaired.[7]

Investigating Leaks

Leaks may be discovered by members of the public, emergency response personnel, or company employees. When members of the public suspect a leak, they generally report them to the company, the police, or the fire department. Unfortunately, members of the public may not recognize or may ignore leak warning signs. Thus, natural gas companies and the natural gas industry work to educate the public regarding leak warning signs and to encourage them to report suspected leaks.

When investigating a reported leak, the first concern is always safety, so the investigating technician's priority is to make the area safe. Any indication of gas in a building or other confined space is considered hazardous. People are evacuated from the area, ignition sources are eliminated, the gas is shut off to the suspected leak area, and the area is ventilated to reduce the gas concentrations to a safe level. Once the area is made safe, the technician begins the leak location process.

Locating Leaks

"Think like gas," the leak detection foreperson tells the new leak technician. What the foreperson means is that the technician must consider the potential sources and how gas migrates. Locating leaks is usually an iterative process of finding the gas migration perimeter and then narrowing that perimeter to pinpoint the leak.

Confined space leaks

For leaks inside buildings, technicians discuss with the person reporting the leak where and when the odor was noticed. Then they use a gas leak detector or leak detection solution to narrow down potential leak locations until they find the leak. Most likely leak sources include fittings and valves, so technicians concentrate on those areas first.

Underground leaks

If underground leaks always vented directly above them, pinpointing their exact location would be easy. However, gas escaping from leaks under roads, driveways, and parking lots may migrate a considerable distance before it comes to the surface. If the leak is underground, technicians commonly use a bar or probe to make a series of vertical holes near the gas line. Industry insiders usually call these vertical holes *bar holes* after the probe or bar used to produce them (fig. 8-11).

Fig. 8–11. Technician using a probe to produce a bar hole
Courtesy: Heath Consultants Incorporated.

Bar holes are driven to a depth at least equal to the depth of the gas line in a pattern around the suspected leak location. Technicians then successively insert the probe of a CGI into the bar holes, observing the reading from each hole. The hole with the highest sustained reading is typically closest to the leak. When underground structures such as water and sewer lines, electrical conduits, and telephone conduits are near the line, they are also tested for gas concentration.

Leak Management

After the leak is located, it is commonly classified according to the following PHMSA definitions:

- *Grade 1.* A leak that represents an existing or probable hazard to persons or property, and requires immediate repair or continuous action until the conditions are no longer hazardous.
- *Grade 2.* A leak that is recognized as nonhazardous at the time of detection but justifies scheduled repair based on probable future hazard.
- *Grade 3.* A leak that is nonhazardous at the time of detection and can be reasonably expected to remain nonhazardous.[8]

It should be noted that the PHMSA classification scheme is consistent with other classification schemes, such as the leak classification guide from ASME.

Technicians consider four factors as they develop the leak classification:

- *Dispersion.* Based on the gas venting perimeter.
- *Location.* The leakage area as compared to the surroundings.
- *Proportion.* The amount of gas being released.
- *Evaluation.* Based on the operator's judgment, experience, and company guidelines.

Taking into consideration these four factors, the final classification is assigned and repair plans are made. Grade 1 leaks present the highest hazard and are repaired immediately. Grade 3 leaks may be so small and nonhazardous they enter an observation period where they are routinely checked to ensure they remain Grade 3. Many companies repair all leaks as a matter of policy.

Summary

- Leaks are a subset of releases.
- Leaks come in a variety of sizes ranging from complete (or nearly complete) separation of the line, commonly referred to as a *rupture*, all the way down to very small leaks, commonly classified as *nonhazardous leaks*.
- As the leak cloud fans out, the amount of methane concentration in the air decreases.
- The lower flammable limit (LFL), also referred to as *lower explosive limit* (*LEL*), of methane is about 5% concentration in air.
- The upper flammable limit (UFL), also referred to as *upper explosive limit* (*UEL*), of methane is about 15%.
- How gas migrates from the leak site to the vent site depends on a number of factors.
- Natural gas companies conduct regular leak surveys using sophisticated instruments to detect system leaks.
- Smell (odor) is the leak detection method most commonly used by the public.
- A soap solution is frequently used to confirm the location of small leaks.
- When using instruments to detect, size, and locate leaks, the three critical factors are venting, instrument, and operator.
- Technicians use FID, OMD, and RMLD instruments to conduct leak surveys.
- Once a leak is detected, a CGI instrument is often used to pinpoint the leak exactly.
- When investigating a reported leak, the first concern is always safety, so the investigating technician's priority is making the area safe.
- Gas escaping from leaks under roads, driveways, and parking lots may migrate a considerable distance before it comes to the surface.
- Technicians commonly use a bar or probe to make a series of bar holes in which they measure gas concentration.
- Once located, the leak is classified.
- There is currently no effective deployed, instrument based, real-time leak detection system for small to medium natural gas local distribution pipelines.

Notes

1. New London Museum, "The Tragic Events of March 18, 1937," *The New London Texas School Explosion*, http://www.newlondonschool.org/index2.html.
2. American Oil and Gas Historical Society, "New London School Explosion," *Petroleum History Almanac*, http://aoghs.org/oil-almanac/new-london-texas-school-explosion.
3. US National Transportation Safety Board, "Pipeline Accident Reports," http://www.ntsb.gov/investigations/AccidentReports/Pages/pipeline.aspx.
4. Pipeline and Hazardous Materials Safety Administration, extracted June 26, 2014 from http://opsweb.phmsa.dot.gov/pipelineforum/facts-and-stats/incidents-and-mileage-report/. Note: This URL now redirects to http://phmsa.dot.gov/pipeline/library/data-stats.
5. US National Transportation Safety Board, "Natural Gas Distribution Line Break and Subsequent Explosion and Fire, Plum Borough, Pennsylvania, March 5, 2008," Pipeline Accident Brief DCA-08-FP-006, NTSB/PAB-08/01 (US NTSB, November 21, 2008), https://app.ntsb.gov/investigations/fulltext/PAB0801.htm.
6. Pipeline and Hazardous Materials Safety Administration, "Leak Detection," in *Guidance Manual for Operators of Small Natural Gas Systems* (PHMSA, US DOT, June 2002), http://phmsa.dot.gov/pv_obj_cache/pv_obj_id_9EE7F8CC855891158EBCECB1F1C9C8EC29CF0900/filename/4%20-%20Guidance%20Manual%20for%20Operators%20of%20Small%20Natural%20Gas%20Systems-2002.pdf.
7. Ibid., IV-5.
8. Ibid., IV-14–IV-16.

chapter 9

Asset Integrity

The strength of a nation derives from the integrity of the home.

—*Confucius*

Introduction

In preparation for his upcoming retirement, the veteran distribution pipeline maintenance director was discussing the job with his replacement. As they climbed into the pickup, the retiring director offered some advice.

"Generally speaking, failures, or leaks, come from one of three broad defect categories. These categories are time-dependent defects, time-independent defects, and stable defects."

As they drove along, the director further explained the defect categories.

"The first category is *time-dependent defects*, which worsen over time. This includes defects due to corrosion or fatigue. The second category is *time-independent defects*, which are defects that happen quickly, possibly from a contractor puncturing a line or from an earthquake, for example. The third category is *stable defects*, which do not get worse over time, like some construction damage or material defects."

"Years ago, we would attempt to analyze what caused leaks, and in what component they normally occurred, trying to understand how to prevent them. But we did not have robust data recording techniques and lacked processing capability to fully understand the condition of the asset, and we could not predict when defects would happen," the director said. "If we can understand where defects are located, how bad they are, and the rate at which they are getting worse, we stand a good chance of fixing them before they fail," added the director.

"Over my 40-year career," the maintenance director continued, "the definition of maintenance has changed from 'fix it when it leaks' to include the following steps:

1. *Diagnose.* To diagnose the problem, we have to understand the condition of the asset.
2. *Predict.* We then determine, based on data and scenarios, potential failure points and rates.
3. *Prevent.* We take actions to prevent failures before they occur.
4. *Repair.* If a failure does occur, we take steps to return the asset to normal operating condition and capacity.

The term *asset integrity* has now largely replaced the term *pipeline and facility maintenance*," the director explained.

"In the LDC business, the process is now generally called *distribution integrity management*. The acronym is DIMP, with the **P** standing for 'plan.' There are two more things," the director added. "Leaks often happen from a combination of causes, not just one, and the most dangerous leak is the one that happens in a confined space where gas accumulates to the explosive range and then ignites. Small leaks out in the open are hardly ever a problem, but even very small leaks in enclosed spaces can be catastrophic," concluded the director.

Risk

The veteran maintenance director was right, but he neglected to explain the critical concept of *risk*. In the health and safety disciplines, risk is generally accepted to depend on the consequences (results) of an event happening and the probability (likelihood) of occurrence.

Risk = Function (Consequences, Probability)

Applying the concept of risk to a familiar topic, travel by car and by plane, the number of people killed by an airplane crash is normally higher than by a car crash, but the probability of an airplane crash per mile traveled is significantly less than that of a car crash, making travel by air less risky (safer) than travel by car. If asked, however, many people would say they perceive air travel as riskier than car travel. Risk increases when consequences or likelihood, or both, increase (fig. 9–1).

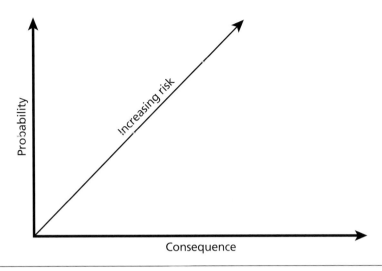

Fig. 9–1. Risk rises with increasing probability and consequences.

Consequences

The magnitude of the consequences of a natural gas leak depends primarily on whether or not the leak ignites. Injuries, fatalities, and property damage may result from gas leaks that ignite. Consequences of leaks that do not ignite may be limited to supply disruptions and release of methane into the atmosphere. Leaks in highly populated areas normally have higher consequences than those in sparsely populated areas, as do those that are captured in confined spaces versus those not in confined spaces.

The severity of the consequences generally varies along the pipeline. A segment running through a densely populated area will have different consequences than a segment in a rural area. Thus, pipelines are divided into linear segments for consideration of consequences.

Probability

W. Kent Muhlbauer, a widely recognized authority on pipeline risk, discusses what he calls the *probability of failure (PoF) triad*. Muhlbauer proposes the basis for this model is examining each failure mechanism (threat) in three parts for the following:

- *Exposure.* Likelihood of force or failure mechanism reaching the pipe when no mitigation is applied.

- *Mitigation.* Actions that keep the force or failure mechanism off the pipe.
- *Resistance.* The system's ability to resist a force or failure mechanism applied to the pipe.[1]

An analogous naming convention is *attack, defense,* and *survivability,* respectively, for these three terms. The evaluation of these three elements for each pipeline segment results in a PoF estimate for that specific segment.

Probability of damage. Even if the pipe is damaged, it may not fail, or it may not fail immediately. The likelihood of damage without immediate failure is determined by using the first two terms, exposure and mitigation.

Probability of failure. Conceptually, whether or not the pipe fails when it is damaged depends on its ability to resist or withstand the damage. Plastic pipe, for example, is less resistant to damage from a backhoe bucket than is steel pipe. On the flip side, plastic pipe is not subject to damage by corrosion, whereas steel pipe is.

With the concept of risk in mind, this text now turns to understanding and preventing failures and concludes with more details about DIMP. Readers wanting to learn more about risk can consult one of Kent's books such as *Pipeline Risk Assessment, The Definitive Approach and Its Role in Risk Management* by W. Kent Muhlbauer, along with the associated websites.[2]

Asset integrity from "repair when it fails" to "prevent and manage the failure"

The pipeline and most other industries have evolved from a mindset of "repair defects when they fail" to a more proactive mindset of "repair defects before they become failures" or "prevent defects from forming." Pipeline personnel now complete the following steps:

- Record asset *attribute data* (such as material of construction, SDR, wall thickness, etc.).
- Collect asset *condition data* (wall loss, cracks, and other damage).
- Access *geospatial data* (population density, seismic activity, roads, and the like).
- Overlay the first two data sets on the geospatial data.
- Analyze the combined data to understand failure patterns, precursors, and consequences of failure.

Turning data into knowledge allows pipeline personnel to locate and repair defects before they grow to the point of failure. Data can be used to prevent defects from forming or to manage them when they are found (fig. 9–2).

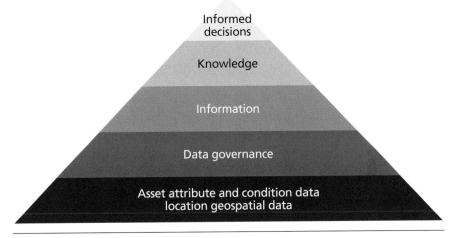

Fig. 9–2. Data to informed decision pyramid

Mining Past Data

Obviously, data-based decisions require data from previous accidents, from previous system inspections, and from original construction. This data is collected and analyzed for failure patterns. Examples of necessary data for prevention and maintenance include the following:

- Cause
- Frequency
- Location on the system
- Material of construction

Incident data

The most robust, and publicly available, database in the world of distribution accidents is maintained by PHMSA. Unfortunately, prior to 2004, this database contained only five broad categories:

- Accidentally caused by operator
- Construction defect

- Operation error
- Corrosion, or damage by outside force
- Other

These broad categories made it difficult to identify the real hazards or mechanisms. For instance, in 2004, the category "damage by outside force" accounted for 61% of the incidents, but lumped together incidents caused by excavation and mechanical damage, incidents caused by natural forces, and incidents caused by "other outside forces," such as vehicles and fires. These causes involve different hazards and different actors, and require different strategies to manage the risks. A full 25% of the incident causes were reported in the category "other," which is effectively a black hole of reporting.[3]

Categories

In 2004 the reporting form was reformatted into seven major categories, each with several subcategories, allowing more detailed data gathering (table 9-1).

Location. Buried lines are affected differently than aboveground meter sets, so the form captures the following location categories:
- Main
- Service line
- Meter set assembly
- Other part
- No part data

Material of construction. Iron and steel corrode, but plastic does not. Accordingly, knowing the material of construction is important. The most common materials of construction are the following:
- Plastic
- Steel
- Copper/other
- Iron

Table 9–1. Causes used for reporting gas distribution incidents on PHMSA form

Corrosion	• External corrosion • Internal corrosion
Natural Causes	• Earth movement • Lightning • Heavy rains/floods • Temperature • High winds
Excavation	• Operator excavation • Third-party excavation
Other Outside Force	• Fire/explosion as the primary cause • Vehicle unrelated to excavation • Previously damaged pipe • Other outside force • Vandalism
Material/Weld	• Material/body of pipe • Material/component • Material/joint • Weld/butt • Weld/fillet • Weld/seam
Equipment/Operation	• Malfunction of control/relief • Threads stripped, broken pipe • Leaking seals • Incorrect operation
Other	• Miscellaneous other • Unknown

Limitations of public data sources

PHMSA maintains a publicly available database of leaks meeting the reporting requirements threshold. Two US Department of Transportation reports examine the PHMSA database. The first, *Safety Incidents on Natural Gas Distribution Systems: Understanding the Hazards*, is authored by Cheryl Trench, Allegro Energy Consulting,[4] and the second is *The State of the National Pipeline Infrastructure*, produced by U. S. Department of Transportation in 2012.[5] The data sources in table 9-2 are not exact comparisons. The data contained in the table are from different time frames, and the definitions of the causes are a bit different. Table 9-2 is, however, useful in understanding at a high level what causes releases, a requirement for developing the integrity plan to prevent them.

Table 9–2. Causes of serious natural gas distribution incidents

Cause	PHMSA Database 1994–2013 (%)	Understanding the Hazards (%)	State of the Infrastructure (%)
Excavation	35	38	18
Other outside force	9	29	22
Other	28	12	29
Equipment/operation	12	7	18
Natural causes	7	7	4
Material/weld	6	5	6
Corrosion	5	3	4

The significantly lower percent for excavation in the *State of the Infrastructure Report* could be the result of using two different time frames. The *State of the Infrastructure Report* looked at only 2008 through 2010 data. Common Ground Alliance, representing industry, governmental, and underground utility excavators, and other groups have dedicated significant resources to publicizing the need to call 811 before digging. This should have resulted in decreased excavation-related incidents. It should be noted that there is a significantly lower percentage of "other" incidents in the *Understanding the Hazards* report. It seems likely that the author of the report took the time to read each incident narrative and appropriately reclassify as many of the incidents in the "other" category as possible into the appropriate category, based on her knowledge of the industry.

The "excavation," "other outside force," and "other" categories comprise about 70% of all serious distribution incidents. However, a word of caution is necessary. These statistics include only serious incidents, and accordingly, the distribution may not be consistent with the distribution when all incidents are considered. Additional analysis of the data is beyond the scope of this text; readers are referred to the two reports and the PHMSA website for additional understanding.

In addition to US data collected by PHMSA, other countries track distribution failures as well. For example, a total of 286 Gas Safety Management Regulations (GSMR) reportable incidents were received in 2012/2013 by the UK Health and Safety Executive (HSE). Of those, third-party damage was the leading cause of incidents. Gas service pipe failure was also increasingly significant and caused as many GSMR reportable incidents as mains failure in the reporting period.[6] Two key initiatives from the UK HSE concern third-party damage and replacement of iron mains.

Company-specific data

While industry-wide statistics serve as a guide, company results can be as unique as fingerprints, depending on the specific material of construction, local excavation activity, and a myriad of other factors. Thus one of the first requirements of company integrity management plans is knowledge of the specific leak history, mapping data, facilities inventory, records of facilities damage, and One Call information. Also needed are incident data, new construction data, records of material or mechanical fitting failures, and the expertise of personnel responsible for the design, construction, operation, and maintenance of the company's systems.

Failure Mechanisms and Forces

Preventing releases involves the following:

- Understanding the hazards, mechanisms, or forces that can cause releases
- Building safety measures into the system:
    - Mitigating measures to prevent the mechanism from reaching the pipe or component
    - Resistance to failure if the forces do reach the pipe or component

The next section of the text contains a brief description of the various failure mechanisms and forces, using categories from table 9–2. This serves as a prelude to the next section, which discusses mitigating measures.

"Excavation" category

Largely comprised of traditional excavation and mechanical damage, this category includes construction activities such as trenching, grading, and other earth-moving activities, and utility installation such as cable, telephone, or water lines. Equipment ranges from a large tracked vehicle to a small backhoe rented by a homeowner to replace a drain line, or even a shovel. The rising popularity of trenchless technologies such as horizontal directional drilling (HDD) introduces a new challenge when it comes to preventing excavation damage. Sometimes the release happens immediately when the line is punctured. In other cases, the line is damaged during excavation but does not immediately fail. Over time the damaged area deteriorates, and the failure happens much later.

Figure 9–3 shows a 2-inch natural gas main that failed five years after the pipe's protective coating had been stripped off by an excavator in the process of installing a sewer line. With the protective coating removed, the steel pipe corroded over time and failed, resulting in a natural gas explosion that destroyed a residence, killing a man and seriously injuring a 4-year-old girl. Two other houses were destroyed, and 11 houses were damaged. Property damage and losses were $1,000,000.[7]

Fig. 9–3. Two-inch failed main
Source: US National Transportation Safety Board (2008).

"Other outside force" category

One large component of this category involves something hitting aboveground facilities such as meter sets. Incidents typically involve an automobile crash and a fire. Some involve a DUI, and some a rollaway vehicle, a riding lawnmower, or snow plow. The subcategory of "vehicle not involved in excavation" is responsible for about three-quarters of the incidents in the larger "other outside force" category.

Figure 9–4 shows a 3/4-inch-diameter polyethylene gas service line leading to a house along with electrical cables installed in the same ditch—a common practice. The plastic line developed a hole when corrosion and subsequent overheating and arcing at a splice in one of the conductors of the triplex electrical service line occurred. While electric and gas lines are commonly installed in the same ditch, they are normally well separated from each other. In this case, inadequate separation between the electrical conductors and the gas service line meant the gas line was so close to the electric cables that the arc melted the plastic line. Gas leaking from the hole accumulated in the basement and ignited.

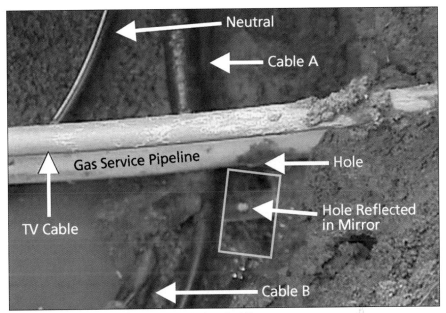

Fig. 9–4. Natural gas service line and electrical lines
Source: US National Transportation Safety Board (1998).

Unfortunately, a family of four was spending their first night in their new home at the time of the explosion. As a result of the accident, the wife was killed, the husband was seriously injured, and the two children received minor injuries. Five other homes and two vehicles were damaged.[8]

"Other" category

"Other" (or "unknown") is a frustrating classification, as it does little to help in understanding and preventing incidents. The PHMSA data list more than 25% of serious incidents as either "unknown cause" or "miscellaneous cause." Reading the narrative accompanying the reports for these cases allows reclassification to the proper cause and reduces this category to about 12%, as shown in the *Understanding the Hazards* column of table 9-2. More than one-half of the "other" incidents remaining after reclassification involved customer piping and appliances, such as furnaces and water heaters, which are beyond the responsibility of the LDC operator. This reclassification effort meant that "other" dropped from more than 25% to about 5%. More importantly, it pointed to the real failure mechanism so protective measures could be developed.

"Equipment/operation" category

Malfunction of control or relief valves, broken pipe, stripped threads, leaking seals, and improper operation are examples of items included in this category by the *Understanding the Hazards* report but not in the other two reports. The PHMSA data list 129 operation-related incidents. Of the 129, 115 are classified as either "other incorrect operation" or "unspecified incorrect operation," making analysis difficult.

"Natural causes" category

The subcategories "earth movement" and "heavy rains/floods" comprise nearly one-half of the serious incidents from natural causes.

"Material/weld" category

Included in this category are items such as failures of either a system component, the pipe itself, connectors, or circumferential or girth welds in the case of steel pipe or fusion in the case of plastic pipe. In other words, material or construction defects. The other two data sources add in the equipment component to this category. Whichever combination is used, materials and equipment are not a large component of significant incidents. While not a large component, any incident can be serious.

Figure 9–5 shows a connection that pulled apart about two years after it was made. The piece of pipe that pulled loose from the compression fitting was unmarked, out-of-specification polyethylene pipe with inadequate wall thickness. When the connection pulled apart, gas leaked from the mechanical coupling and migrated into a house, resulting in an explosion and fire. One person suffered fatal injuries, and five other people, including one utility employee and one firefighter, were hospitalized as a result of the explosion. Two adjacent homes had severe damage, and several homes suffered minor damage.[9]

This incident points out one of the dilemmas in assigning cause. The plastic pipe was clearly defective. However, the unmarked pipe should never have been installed, and someone had to let it into the system. Failures often result from a combination of factors.

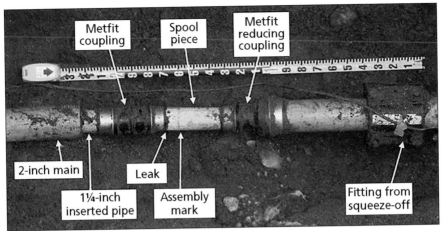

Fig. 9–5. Two-inch natural gas main with failed connection
Source: US National Transportation Safety Board (2010).

"Corrosion" category

Iron corrodes (rusts) when unprotected. Either the pipe's outer wall (*external corrosion*) or its inner wall (*internal corrosion*) loses iron, reducing the pipe's wall thickness and its ability to contain pressure. (Steel is mainly iron with small percentage of other elements.) Different time periods provide different results, but over time, the combined internal and external corrosion comprise less than 5% of distribution significant accidents.

Examining the data reveals that even though wrought and cast iron comprise a very small portion of local distribution systems, they have a high share of the corrosion leaks. Companies have proactive cast iron replacement programs to remove this potential threat from their systems.

Conclusions

While a comprehensive analysis is beyond the scope of this text, definite conclusions can be drawn from the data and past studies.

- Externally inflicted damage, whether during excavation activities or from vehicle crashes, is the leading cause of serious incidents.
- Beyond externally inflicted damage, there are a variety of factors causing leaks, including fires, which can compromise natural gas systems, causing them to leak and fueling the fire.
- Wrought and cast iron should be systematically replaced.

- As with most accidents, there is seldom one single cause. Several things normally go wrong at the same time.

Preventing Releases

Preventing releases seems simple: just prevent the forces that cause damage from reaching the pipe or component, or if the force does reach the pipe or component, build resistance into the system to keep it from failing.

These simple concepts have grown into the science of integrity management. Local distribution (and gas and oil transmission) integrity professionals divide the topic into three factors:

- *Exposure.* Likelihood of force or failure mechanism reaching the pipe when no mitigation is applied. Areas of high development growth and frequent excavations have more exposure than a subdivision fully built out, for example. However, there is still exposure in areas with less-recent development. For instance, external forces such as a hurricane or tornado could damage or destroy privacy fences, necessitating their rebuilding and the attendant installation of new posts.
- *Mitigation.* Actions that keep the force or failure mechanism off the pipe. Dig Safely and Call 811 are mitigation measures aimed at educating excavators to increase their awareness. Coating is a mitigation measure designed to keep steel pipe from corroding.
- *Resistance.* The system's ability to resist a force or failure mechanism applied to the pipe or components. Supports and anchors can be added to provide resistance to damage from natural causes such as ground movement.[10]

Each of the various failure categories or causes may require different means to reduce exposure, improve mitigation techniques, or increase system resistance to failure.

Excavation incidents

In the early to mid-1990s, both the local distribution and the transmission industries undertook initiatives focused on improving the understanding of what causes failures and the frequency of those failures. Excavation damage emerged as a major threat. In 1998, with the support and encouragement of the natural gas industry, Congress established a national "Call Before You Dig" safety program, known as *One Call*.

One Call is aimed at developing a variety of best practice procedures to prevent excavation damage to underground facilities.

In 2000, a coalition of underground excavation industry stakeholders came together to form the Common Ground Alliance (CGA). CGA represents a continuation of the damage prevention efforts embodied by the Common Ground study. Sponsored by the US Department of Transportation and completed in 1999, this study represents the collaborative work of 160 industry professionals who identified best practices relating to damage prevention.

The CGA provides a forum where stakeholders can share information and perspectives and work together on all aspects of damage prevention issues. CGA is not limited to pipelines. It works across underground excavation industry stakeholders and regulators to produce stronger, more effective results through partnership, collaboration, and the pursuit of common goals in damage prevention. Its mission is "to enhance safety and protect underground facilities by:

- Identifying and disseminating the stakeholder best practices;
- Developing and conducting public awareness and education programs;
- Sharing and disseminating damage prevention tools and technology; and
- Serving as the premier resource for damage and one call center data collection, analysis and dissemination."[11]

The following sections draw heavily on a report by CGA, *Best Practices*.[12]

Planning and design. Knowing the locations of pipelines, and designing construction projects to minimize the need to dig around pipelines, is one of the best ways to avoid pipeline accidents caused by equipment. When a project involves excavation, the designers attempt to locate all underground utilities in the area, which are then clearly shown on project plans. Designers and utility operators should meet to determine how best to work around the lines, including excavators early in the planning process to achieve the goal of digging safely.

One Call centers. One Call centers are clearinghouses for excavators to learn about all underground utilities in a specific geographic area. Designers and excavators should contact the center giving them the location of their projects. (It is equally important for homeowners to call their local utilities or One Call center before beginning a "backyard project.") The centers then consult maps and databases and contact the companies owning the underground facilities close to the excavation location. Most states in the United States have one or more One Call centers. Fortunately, the Common Ground Alliance has fostered improved cooperation between various involved parties and the offices of the area's jurisdictions.

Locating and marking. If a pipeline is contacted by a One Call center, it checks its records to see how close the excavation is to its lines. If it is close enough, the pipeline company sends someone out to locate and place markers along the line.

Pipeline employees or contractors use their maps, line finders, and probes to locate and stake the exact location of pipelines prior to the start of excavation projects (fig. 9–6).

Fig. 9–6. A technician points to a map on the laptop showing the locations of the main and service lines.

For particularly confined excavation sites or where the location of lines is uncertain, lines are hand dug to find the exact location prior to mechanical excavation. Hand digging costs more than mechanical excavation, but it can avoid costly accidents. Many pipeline operators require the presence of their own inspectors during any excavation work near their lines. Excavators and pipeline operators normally work together to plan their activities and allow proper inspection and monitoring.

Excavation. Prior to excavation, representatives from the excavator and pipeline operator meet to agree on the excavation plan and confirm line locations. A tolerance zone, equal to the width of the pipeline plus a short distance on each side, is established. (CGA's *Best Practices* suggests at least 18 inches on each side.) No mechanical excavation is allowed within this tolerance zone. Spotters, the construction company employees assigned to monitor the excavation, warn the equipment operator of dangers not easily seen. Pipeline operators monitor their own contractors digging near the line as vigilantly as they monitor other excavators.

Maps. Accurate maps sometimes make physically locating the line during initial project design unnecessary. Supplying pipeline location coordinates directly to project designers for inclusion on the drawings makes their work easier and improves safety. Even with the most accurate maps, pipeline operators, designers, and excavators often visit the proposed construction site prior to final design approval to verify locations.

Regulations and compliance. Appropriate regulations and compliance are important factors in avoiding equipment damage. Excavation regulations tend to vary from country to country, and often within a country as the result of differing requirements by local jurisdictions. In general, regulations requiring One Call systems and establishing tough fines for excavators who do not use the centers or who damage underground utilities tend to reduce pipeline accidents.

Public education. A myriad of stakeholders (landowners, residents, excavators, utility operators, legislators, regulators, and emergency response personnel) ought to be knowledgeable about pipelines in order to reduce the risk of pipelines accidents. Pipeline companies conduct education programs, informing those living and working around the line about how to protect it and how to recognize and respond to emergencies. Informed landowners are a valuable first line of defense for the pipeline operators. Beyond that, in the event an accident happens, trained emergency response personnel are essential.

Monitoring the route. Pipeline operators regularly patrol the route from the air as well as from the ground. These patrols look for excavation near the line and indications, such as stakes, markers, and vegetation clearing, that excavation is likely. At the same time, the patrols look for small releases, encroachments, erosion, exposed pipe, marker condition, overgrown vegetation, earth movements, and general right-of-way condition. Technologies are improving route monitoring, including surveillance by satellite, sonar mounted on airplanes, and fiber optic cables installed along the right-of-way.

Reporting and evaluation. The final step in reducing equipment damage is reporting and analyzing data about previous incidents to understand the root causes, and to take actions to reduce the likelihood of future damage. Collecting the information in a consistent format and providing it to one organization for analysis is the best way to learn about damage and how to avoid it. Industry associations and governmental agencies normally pick up this task. Whoever does it should include representatives from all stakeholders in "two-way" communications. Stakeholders can be educated about equipment damage and their role in preventing it. They can also offer valuable insight into improvement techniques.

Interested readers can learn more at the CGA website (commongroundalliance.com).

Other outside force damage

Two primary sources of outside force damage are vehicles crashing into facilities, primarily meter sets, and feeding fires started by other means. With this in mind, LDCs have installed protective barriers to keep vehicles from reaching the meter sets and aboveground valves (fig. 9–7).

Fig. 9–7. Barrier installed to keep vehicles from striking a district regulator assembly along a major road

Federal regulations require installation of excess flow valves (EFVs) on all service lines installed since June 2008, as well as in some other areas where the potential for rupture is significant. Most LDCs will also install excess flow valves in service lines that predate June 2008, if requested, at the customer's expense. EFVs close automatically when flow reaches a predetermined level, shutting off flow of gas so fires are not fed.

The other 40%

Completely eliminating damage from excavation activities, car crashes, and fires that start from other causes and compromise the gas system would eliminate about 60% of serious incidents. That leaves the other 40% comprised of a variety of causes.

Other. Taking actions to prevent incidents of unknown cause is nearly impossible, so the industry and regulators are working hard to improve reporting and better understand this category. As mentioned earlier, many incidents in the "other" category do not involve the LDC system at all. They happen inside the house to piping and appliances.

Equipment/operations/materials. Two categories are lumped together in this section. Incident prevention in both can be simplified as excellence of materials, equipment, construction, maintenance, and operations. There is nothing magical here, just old-fashioned operations excellence. Some of the components of operations excellence involve the following:
- Hiring
- Training
- Coaching
- Communicating
- Procedures
- Standards
- Specifications
- Inspection
- Testing
- Calibration

This list goes on but involves doing the job right, all the time, every time.

Corrosion. Veteran pipeliners from the transmission side of the industry would find the low rate of serious incidents due to corrosion in local distribution systems shocking. On further thought, though, it is not that surprising. First, plastic, which does not corrode, comprises a great deal of the systems. Second, cast iron and steel pipe walls are much thicker than required to contain the relatively low distribution pressures, leaving a significant corrosion allowance. Finally, corrosion protection has significantly advanced as a science with effective external coatings and cathodic protection. As discussed previously, corrosion is current flow. When it flows in the "wrong" direction, steel rusts. Forcing current to flow in the "correct" direction makes the pipe the cathode (as opposed to the anode), and thus it does not rust. Keeping current from flowing at all via coatings also prevents rust.

However, over time, unprotected iron rusts. Operators have replaced, and continue to replace, their cast iron and bare steel pipes (fig. 9-8).

Fig. 9–8. Plastic service line installed inside an existing steel service line

Integrity Management Plans

Integrity management plans (IMPs) provide the framework for assessing and mitigating risks. IMPs are not unique to the local distribution industry; many industries use them. Every industry is different, and there are often specific differences between companies, so IMPs vary as appropriate depending on the particular assets and the identified risk. Figure 9-9 shows the process flow for a generic integrity management plan.

Chapter 9 Asset Integrity   243

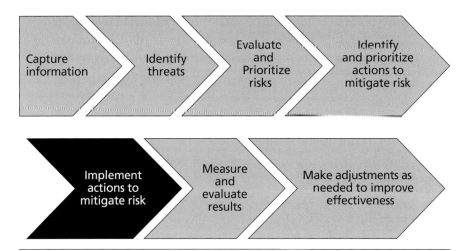

Fig. 9–9. Integrity management plan flow chart

Plans are written, reviewed, updated, and improved as lessons are learned. While each individual step or element is important, the "implement actions to mitigate risk" step produces results and improves performance. The four steps before implementing actions ensure the actions are the most efficient and effective, and the two steps following provide data for improving the process. Examples of actions to mitigate risks include the following:

- On a specified frequency, walk the course of pipelines that have experienced problems in the past to look for signs of damage and to smell for gas.
- Monitor more frequently any portions of the system experiencing frequent leakage.
- Coat and cathodically protect all areas of metallic pipe experiencing active corrosion.
- If a section of the pipeline is leaking at a high rate, or is found to be deteriorating due to corrosion, replace the section of pipe.
- Conduct more frequent patrols to identify conditions that may adversely affect pipe or components, especially following lightning storms, tornadoes, earthquakes, landslides, flood-induced erosion, or high winds leading to uprooting of nearby trees.
- Implement a damage prevention program, including the following elements:

- A means of receiving and recording notification of planned excavation activities.
- Requirements to locate and mark the pipe in areas where buried piping exists and excavation is planned.
- Provision for actual notification of persons who indicate intent to excavate in areas where buried pipe is located, with a description of the temporary markings and how to identify them.
- Provision for inspection of pipelines during and after excavation if there is reason to believe they could be damaged.

- Implement measures to reduce the opportunities for vandalism to effect the safe operation of the system.
- Replace small diameter cast iron or ductile iron pipe that is not adequately supported.
- Replace brittle plastic pipe or other materials unsuitable for gas service based on its leak history.
- Implement the recommended actions in any notice received from a pipe/fitting manufacturer regarding material defects.
- Replace any component in the system that has experienced a number of failures or is failing.
- Provide additional integrity training to the workforce.[13]

In early 2009, the Northeast Gas Association (NGA) and the Southern Gas Association (SGA) announced the development of a joint collaborative to develop a framework document for distribution integrity management plans. The framework document serves as an outline. From the outline, distribution companies build customized distribution integrity management plans based on their systems and needs. Various software vendors provide databases to capture knowledge and automate risk assessments and prioritization, from which the company determines their actions and builds their plans.

Determining Asset Condition

Transmission line integrity engineers extensively use internal line inspection (ILI) tools to find wall loss and cracks in steel lines. However, the small diameters, low pressures, multiple connection points, and 90° bends in gas distribution systems means ILI tools cannot be pushed through most mains, and essentially none of the service lines. Even if they could, current ILI technologies do not work on plastic. LDC integrity engineers do insert cameras and other devices into the lines when they suspect defects.

Defects that are not leaking are most often found by looking for them in areas where history says they are most likely to exist. For example, materials that are susceptible to corrosion or other material failure, as well as soil environment, pressure, how the pipe has been maintained, its leak history, and even depth of burial and frost depth, owing to potential frost heave problems. Records are analyzed to determine these areas of most likely defects and LDCs develop systematic programs to look for defects in these areas.

Defects and Failures

Defects that have become failures, or breached lines, are most often found either during periodic leak surveys, by smell, or by the person who punctured the pipe (fig. 9–10).

Fig. 9–10. Service line leak "found" by the fence contractor who created it digging a post hole after Hurricane Rita.

Repairing Defects and Failures

Once a defect is discovered, it is repaired if it presents an immediate hazardous condition or is placed on a prioritized list for future repair if it is not hazardous. Often, nonhazardous leaks (those with very low concentrations of gas in the air) are not put on a list for repair but are quickly repaired by the technician who locates the leak as they are already on-site and often have the leak area exposed.

Repairing usually means replacing the segment containing the defect. First, the technician stops flow and evacuates gas from the line. Small diameter plastic lines are normally *squeezed off*, i.e., a clamp squeezes the two sides of the pipe together, stopping flow (fig. 9–11).

Fig. 9–11. Clamp squeezes off flow.

When it is safe, the defect is cut out and a new segment is installed (fig. 9–12).

Larger diameter plastic pipes and any size steel lines are not squeezed off. Instead, the line may be *hot tapped* and *plugged*, that is, two tees are welded onto the pipe, one on each side of the defect. Next, flanges are welded onto the tees, and, one at a time, a tapping machine is bolted to the flanges. The tapping machine cuts a hole in the pipe through each of the tees. The pieces of pipe that were cut out are withdrawn, and a plug is installed through the tees on each side of the defect. The two plugs, one on each side of the repair site, stop flow, and the defective section is removed. Finally, a new section is welded into place, and the plugs are removed and blind flanges are installed on the tees to which the tapping machine was bolted. Figure 9–13 shows repair by hot tapping in progress.

Fig. 9–12. New segment installed using compression fittings. Note the bubbles in the ditch from the soap solution used to check for a gas-tight fit. Also note the new piece of tracer wire spliced in.

Fig. 9–13. Repair by hot tapping. Note the two plugs, one on either side of the cutout. The welder is preparing the pipe for welding, while the helper holds the replacement section in the upper portion of the figure.

Summary

- The definition of maintenance has changed from "fix it when it leaks" to "diagnose, predict, prevent, and repair, if needed."
- Turning data into knowledge allows defects to be located and repaired before they grow to the point of failure. It is also useful to prevent them from forming or to manage them when they do occur.
- One of the first requirements of company integrity management plans is to gather data. IMPs require knowledge of specific leak history, mapping data, facilities inventory, records of facilities damage, One Call information, incident data, new construction data, and records of material or mechanical fitting failures. In addition, IMPs rely on the expertise of personnel responsible for the design, construction, and operation and maintenance of the company's systems.
- Externally inflicted damage, whether during excavation activities or from vehicle crashes, is the leading cause of serious incidents.
- Beyond externally inflicted damage, there are a variety of factors causing leaks. One such factor is fires that compromise natural gas systems and cause them to leak, fueling the fire.
- Wrought and cast iron pipelines should be systematically replaced.
- As with most accidents, there is seldom only one cause. There are often several contributing causes.
- Local distribution integrity professionals divide the likelihood of a release into exposure, mitigation, and resistance.
- Each of the various failure categories or causes may require different means to reduce exposure, improve mitigation techniques, or increase system resistance to failure.
- The CGA provides a forum for stakeholders to share information and perspectives and to work together on all aspects of damage prevention issues.
- Integrity management plans provide the framework for assessing and mitigating risks.
- Defects that are not leaking are most often found by looking for them in areas where history says they are most likely to exist.

- Once a defect is discovered, it is repaired if it presents an immediate hazardous condition or is placed on a prioritized list for future repair if it is not hazardous.
- Repairing usually means replacing the segment containing the defect.

Notes

1. W. Kent Muhlbauer, *Pipeline Risk Assessment: The Definitive Approach and Its Role in Risk Management* (Expert Publishing, 2015).
2. Ibid. Further information is available at http://www.pipelinerisk.com/ and http://www.pipelinerisk.net/.
3. Cheryl Trench, *Safety Incidents on Natural Gas Distribution Systems: Understanding the Hazards*, report prepared for the Office of Pipeline Safety, US Department of Transportation (New York: Allegro Energy Consulting, April 2005), 18.
4. Ibid.
5. US Department of Transportation, "The State of the National Pipeline Infrastructure," Secretaries Infrastructure Report (Washington, DC, 2012), https://opsweb.phmsa.dot.gov/pipelineforum/docs/Secretarys%20 Infrastructure%20Report_Revised%20per%20PHC_103111.pdf.
6. Health and Safety Executive (HSE), *Major Hazard Safety Performance Indicators in Great Britain's Onshore Gas and Pipelines Industry*, Annual Report 2012/13, Hazardous Installations Directorate, Gas & Pipelines Unit (HSE), 15.
7. US National Transportation Safety Board, "Natural Gas Distribution Line Break and Subsequent Explosion and Fire, Plum Borough, Pennsylvania, March 5, 2008," Pipeline Accident Brief DCA-08-FP-006, NTSB/PAB-08/01 (US NTSB, November 21, 2008), https://app.ntsb.gov/investigations/fulltext/PAB0801.htm.
8. US National Transportation Safety Board, "Natural Gas Explosion and Fire in South Riding, Virginia, July 7, 1998," Pipeline Accident Report DCA-00M-P006, PB2001-916501 NTSB/PAR-01/01 (Washington, DC: US NTSB, 1998), http://pstrust.org/docs/ntsb_doc8.pdf.
9. US National Transportation Safety Board, "Pipeline Accident Brief," Accident DCA09FP003. NTSB/PAB-10/01 (Washington, DC: US NTSB, May 18, 2010), http://www.ntsb.gov/investigations/AccidentReports/Reports/PAB1001.pdf.
10. W. Kent Muhlbauer, *Enhanced Pipeline Risk Assessment*, Part 1, "Probability of Failure Assessments," Rev. 4. (WKM Consultancy, 2008), 9, http://www.dnvusa.com/Binaries/Enhanced%20Pipeline%20RA_Part1_Muhlbauer_tcm153-573272.pdf.
11. Common Ground Alliance, "CGA Mission," http://commongroundalliance.com/about-us.

12. Common Ground Alliance, *Best Practices 12.0: Putting Great Ideas in Motion* (Arlington, VA: CGA, March 2015). The latest version is available online at http://commongroundalliance.com/best-practices-guide.
13. Pipeline Hazardous Materials and Safety Administration, "Model IM Plan for Master Meter and Small LPG Operators," Office of Pipeline Safety, US Department of Transportation (Washington, DC: PHMSA), 9–10.

chapter 10

Control Systems and SCADA

The first rule of any technology used in a business is that automation applied to an efficient operation will magnify the efficiency.
The second is that automation applied to an inefficient operation will magnify the inefficiency.

—Bill Gates

    As the chilly dawn breaks, home and office thermostats dial up, igniting furnaces; water heaters light up as people shower; and a plethora of other appliances go into action, all demanding natural gas. Pressure sensors located at selected points on the pipeline grid detect falling pressures, and the system responds accordingly. Some sensors and transmitters are electronic, while others are pneumatic. Whatever the case, they either control valves directly or feed signals to smart devices located along the line or in pressure control or metering stations.

    Depending on how they are designed and programmed, sometimes the local station smart device makes an independent decision. Often the local station smart device sends the information on to central control room servers via dedicated phone lines, radio signal, satellites, or Internet connections. Display screens grouped on control consoles display operating parameters, such as pressures, flow rates, temperatures, and gas quality. Displayed on the same (or other) screens are device status data concerning valve position (open, closed, or in transition), odorant injection rate, control valve pressure set points, and other operating data. The operating consoles are often integrated with, or at least contain, geographical (or geospatial) information systems (GIS). Most people think of GIS as electronic maps, but the systems commonly contain much more information about the pipeline grid and the geography it covers. Figure 10–1 shows a typical central control room.

Fig. 10–1. Typical LDC central control room

Observing the dropping pressures, the control room operator decides to begin receiving gas from the second transmission pipeline serving the city gate. The operator positions the cursor over the valve icon on the display. The valve is currently closed, so the operator clicks on the icon bringing up a dialogue box. The operator clicks on Open and confirms the command. Flow starts, increasing pressure. Satisfied, the operator moves on to other tasks, trusting the control system to monitor operations (fig. 10–2).

Opening the valve starts a whole string of other events. As gas flows from the transmission line, pressure drops in the line. To compensate for this dropping pressure, operators in the transmission control room send commands to compressors at 10 stations along the line, causing them to start. Upstream pressure falls. Control systems respond, withdrawing gas from storage.

Returning to the console from a quick trip to the galley, where lunch was heated in the microwave, the operator receives a call from one of the company's maintenance crews. They are responding to an emergency call from a landowner who punctured a 2-inch distribution main while operating a rented trencher. The crew wants to know about the surrounding feed to the other houses on the line. If there is a one-way feed, the houses downstream from the puncture site will be out of service while the crew repairs the leak. If there is two-way service, they can simply squeeze off on either side of the leak while the other houses continue receiving gas. The operator checks the maps and informs the crew that there is two-way feed; he then returns to the galley to reheat lunch. A cold front blows in later that afternoon, so the control room operator calls one of the connecting transmission pipelines, requesting more pressure. Control systems, the topic of this chapter, make it all possible.

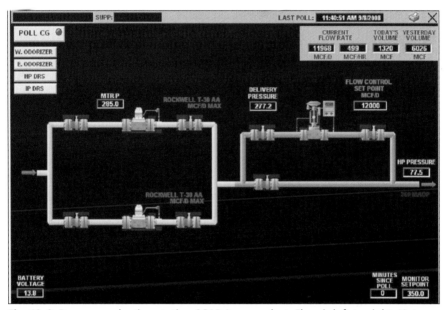

Fig. 10–2. Pressure reduction station SCADA screenshot. Flow is left to right. Note the higher pressures to the left and lower pressures to the right.

Control Systems

Instruments, transmitters, smart devices, communication systems, software, hardware, firmware, and many other components combine to control pressures and flow rates in the line. Across the spectrum from communities of less than 1,000 customers to major cities like London, no two control schemes or control rooms are exactly the same. Some are attended around the clock, providing engineering and dispatch services in addition to control. Others operate essentially unattended.

While the control schemes might be quite different, the goal is the same: to deliver natural gas to each customer—homes, schools, hospitals, businesses and other establishments—at the designated pressure and energy content, safely, reliably, and efficiently, and in an environmentally responsible manner. Each control system component must work together with the others all the way through the system to the final pressure control device.

One of the simplest control devices is the diaphragm valve located at most home meters. Gas pushes against one side of a membrane (diaphragm), and a spring pushes against the other side. As gas pressure builds, the spring is compressed, and the rod connected to the diaphragm moves the valve to pinch down on the gas stream, reducing the pressure into the house. When gas pressure falls, the opposite happens. In this case the feedback loop is very simple and reliable. The pressure at which the valve begins to move open or closed is called its *set point*, or the pressure the valve is "programmed" (through adjustments to spring compression) to maintain.

Working back upstream from the home delivery meter, pressure and gas flow through several hundred delivery control valves, impacting the pressure at the next upstream point, where valves open and close accordingly. The pressure at the next upstream regulating point may also be a simple gas-operated control valve. At this point the control valve is likely to be connected in series with another valve just like it. One of the two valves is the primary control valve. The second valve monitors the first valve, serving as a safety backup. The set point of the monitor valve is just slightly higher than the set point of the primary control valve. If the primary pressure valve fails, the monitor valve ensures pressures remain at safe levels (fig. 10–3).

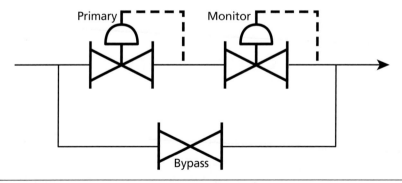

Fig. 10–3. Pressure regulator station schematic. The monitor valve serves as a backup to the primary control valve.

Simple control systems like the one in figure 10-3 are only controlled locally by the pressure in the line and do not report directly into any smart devices. Major city gate stations receiving from multiple transmission lines and distributing into several distribution mains have considerably more complex control schemes. Control is often facilitated by local station and centralized SCADA systems.

SCADA

SCADA is the acronym for supervisory control and data acquisition. SCADA would have been more aptly named DAASC, for data acquisition and supervisory control, since acquisition almost always has to happen before control. But SCADA has an easy, two-syllable ring to it, as opposed to DAASC, and thus has become entrenched.

SCADA systems are not unique to local distribution pipelines. They monitor and control processes in chemical plants, oil refineries, water distribution systems, auto manufacturing, electrical transmission, and other industries. SCADA suppliers build basic systems and then configure (customize) them to suit the idiosyncrasies of each industry sector.

Pipelines have two peculiarities that set their SCADA systems apart from most other industries. First, pipelines cover large geographic areas, such as entire cities, for example. When something unexpected happens, control room operators cannot just "walk over and take a look" as they can in a plant. The geographic location points on the pipeline, and the event the control room operators are looking at, may be miles apart. The distance gives a level of uncertainty and complexity that makes the communication part of pipeline central SCADA critical to successful operation.

Second, pipelines sometimes face regulations and standards imposed by geographically and jurisdictionally overlapping regulatory bodies. Given these factors, the control system philosophies of any two local distribution systems may be quite different.

What SCADA includes varies. Some say it is only the supervisory software monitoring and controlling the pipeline, but not all the other instruments, gauges, transmitters, switches, actuators, computers, displays, databases, software, and communications necessary for the SCADA system to do its job. Others say it includes the full range of instruments, from the field location to computer displays in the central control room. Whatever the scope, all the pieces work to monitor and control the pipeline. Even though the purists would object, the terms *SCADA* and *controls* are used interchangeably in this text.

Local station SCADA

At the local station level, a city gate for example, various devices monitor process variables such as pressures, temperatures, flow rates, and the like, and report those variables into one or more smart devices. These smart devices consider the process variables and make adjustments to station operations according to their programs (fig. 10–4).

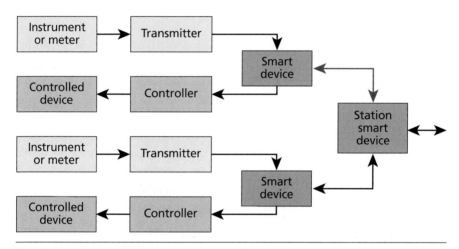

Fig. 10–4. Instruments monitor process variables and transmit those variables to station smart devices.

Centralized SCADA

Process variables from local stations are consolidated at central control rooms. The schematic in figure 10-5 shows the basic hardware and communication links of a typical centralized SCADA system. Each piece of hardware has the appropriate operating systems, application software, and databases.

Chapter 10 Control Systems and SCADA 257

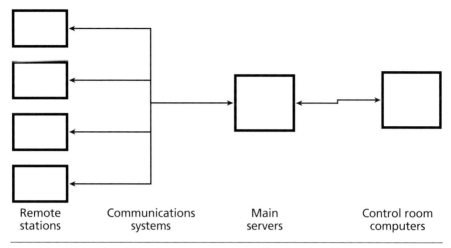

Fig. 10–5. Basic centralized SCADA process flow

Design and Control

Step one in designing a control system should be defining the intent. What is it supposed to do? Who in the organization will be responsible for doing it? With control systems, as with many things—marriage, back surgery, political appointments—the ideal seldom happens. Part of the reason is that very few control systems are designed completely from scratch. Many are burdened with systems, applications, and processes already in place. Control system designers are blessed with (or sometimes burdened by) their unique sets of experiences, viewpoints, and needs.

Human versus machine decision making

Gas distribution pipeline systems started well before the advent of computers. Obviously, humans made all the decisions. But how much of the decision making should be done automatically without human intervention, and how much should be left to human control room operators? This is one of the first issues control system designers must address.

Machines can make decisions quickly and consistently based on the information given to them. They do not "think" broadly about the information. Humans can introduce intuition, judgment, and feelings. That, of course, is sometimes a positive and sometimes a negative.

Some operating decisions are simple and routine and easily made by computers. Other decisions are more complex, such as what actions to take when a leak is suspected or how to respond to upset conditions. Like most other things, decision philosophies vary among companies and among individual control system engineers.

Before moving on, it is important to point out that machine decision making does not always involve a computer. It often involves simple machines such as a pressure regulator located along the line. Components inside the regulator move to throttle out more or less pressure. The SCADA system may monitor the pressures, but the regulator makes pressure adjustments automatically.

Local or central control

After deciding who makes the decisions, the next step is to determine where those decisions will be made. The answers have to be at a local regulator site, in a local station, in a central control room, or some combination of the three. Some LDCs decide not to have central control rooms, and some, depending primarily on size and complexity, do not have full-time dedicated control room staff.

Whatever the situation, local control and central control each have advantages and disadvantages. Central control better enables understanding of what is happening on the entire pipeline grid, thereby facilitating system optimization. One central control room operator making the decisions instead of several local control room operators also means less verbal communication between control room operators. One control room operator can take into account all variables on the grid. But central control requires more electronic communications between local stations and the central control room, since the central control room operator needs to see what is happening at each station. A central control philosophy normally has lower personnel costs, but higher equipment and communication costs, than a local operations philosophy.

At the same time, in the unlikely event communications are lost, control room operators are unable to perform. Robust backup systems make this situation rare, but when it happens, the control room operator cannot know what is happening on the grid and cannot send control commands to change operations. Absent anything else, the pipeline grid continues to operate in the same mode as when communications were lost. In most instances, this is acceptable given the operating pressures and safeguards. In critical situations, field operators are sent to operate the stations to ensure safe operations.

Manual or automated

Early pipelines started out all manually operated. Pressure regulators controlled the system, and valves were opened and closed by hand as needed. Now it is possible to automate most of the functions. Highly automated locations are still sometimes operated locally by an operator walking out and pushing a button. How much and which pieces of equipment to automate are important philosophical decisions that impact overall control design.

Communications

Hard-wired connections, phone lines, cellular or wireless technologies, microwaves, radio, fiber optic, and satellite are all used to move data between field devices and between central control rooms and local control locations. The choice depends on availability, economics, and sometimes personal preference. Often a primary and secondary backup system is used. Radio communication with telephone dial backup is one example of primary communications with a backup for redundancy and reliability (fig. 10–6).

Fig. 10–6. City gate station. Note the radio antenna on the right of the control building.

Human-machine interface

The *human-machine interface* (*HMI*) is the link between the control room operator and the control system, providing visualization and access to controls. It includes computers, screens, keyboards, and mouse devices, usually located on the control room operator's console (fig. 10–7).

Fig. 10–7. Natural gas control room HMI

Control room operators can choose the specific information they want on each screen based on what is happening at the time. They can look at displays, screens, or screenshots, all three of which sound like the same thing. However, there are distinctions:

- *Displays* are the physical devices on which information is displayed.
- *Screens* are the pages of information visible on the display.
- *Screenshots* are pictures of the screen at a point in time.

Screens are the visual information link between the control room operator and pipeline. Easy-to-read screens with a high level of *information density* (lots of information on one screen) help the operator quickly grasp current operating conditions to effectively monitor and control the pipeline. Operators can display a variety of screens but typically have a few critical favorites. One of the favorites is normally a system overview (fig. 10–8).

The navigation screen shown in figure 10–9 allows quick access to other more detailed screens.

Chapter 10 Control Systems and SCADA 261

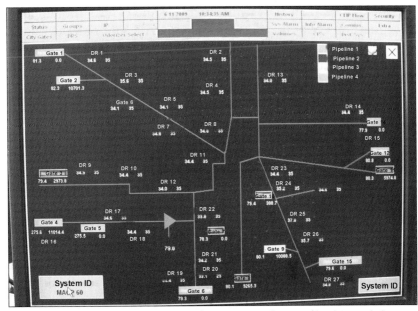

Fig. 10–8. System overview. This screen shows connections to four transmission pipelines, along with 15 city gates and 210 district regulators. The operator can see at a glance the pressures at each.

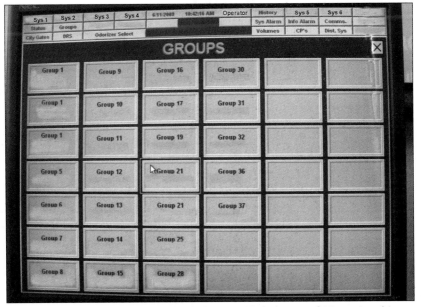

Fig. 10–9. Group navigation page. Operators can click on any of 32 geographic groups to drill down for more detail. Along the top of the screen are other navigation buttons.

If a field or maintenance technician calls the control room operator with a question about Group 8, for example, the control room operator clicks on the Group 8 tab to go down one more level of detail (fig. 10–10).

Ready access to operating parameters makes for efficient operations. The next level of information is the city gate station (fig. 10–11).

Screens often contain the option to navigate to other information. Clicking on the odorizer button in the upper left of the screen shown in figure 10–11 reveals still another level of information (fig. 10–12).

To adjust pressures, a control room operator clicks the mouse to change the pressure setting and open and close valves. Over the years, the HMI has come a long way from the old *mimic boards*, station schematics on the wall indicating the status of valves with lights.

Screens are designed and built as part of the SCADA installation process. Over time the screens are modified and updated as the equipment changes. If a new valve is added to a station, it must be added to the display and the proper communication path installed (fig. 10–13).

STATION	SUPPLIER	CURRENT MCF/D	TODAYS MCF	PRESSURE	SET PT	LAST POLL TIME
Group - 8 Area 1						
City Gates						
Station 1	PL 1	3094.1	189.4	39.1	40	10:31:56 AM 6/11/2009
Station 2	PL 1	0.0	0.0	40.7	35	10:34:27 AM 6/11/2009
Station 3	PL 2	1539.8	97.9	120.1		10:31:31 AM 6/11/2009
Station 4	PL 3	0.0	0.0	38.3	35	10:34:22 AM 6/11/2009
Station 5	PL 1	0.0	0.0	37.3	35	10:31:03 AM 6/11/2009
DRS'S						
Station 1	DRS			39.4	35	10:31:31 AM 6/11/2009

Fig. 10–10. Area page. This screen contains a summary of five city gates' and one district regulator's operating parameters.

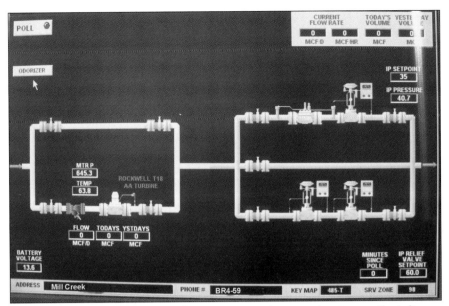

Fig. 10–11. Close up of city gate. The pressure regulators and meters on the left are owned by the transmission pipeline; the pressure regulators on the right by the LDC.

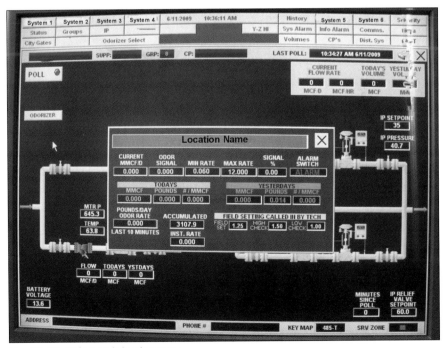

Fig. 10–12. Odorant injection rate

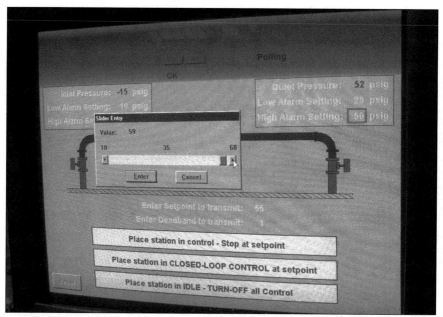

Fig. 10–13. Pressure monitor screen. Low- and high-pressure values are shown. The operator enters a new value into the dialogue box to increase or decrease the pressure settings.

Adding a new city gate station or interconnect requires building new screens and rebuilding existing screens to accommodate the station.

From the main SCADA server, the computers driving the HMI receive data that they turn into information and display for the control room operators. Conversely when control room operators decide to issue a command to open a valve or change pressure settings, the HMI receives the instructions through a switch, keyboard, or mouse and sends it on for action. In addition to displaying information and processing commands, the HMI may accept views from other applications and have graphing and trending capabilities.

Real-time operating data

Real-time describes what is happening on the grid in the present. Strictly speaking, it actually is within a few seconds of the present, or whatever time it takes for a signal to travel from a remote location to a display. Operating data are the important variables associated with pipeline operations—pressures, flow rates, position of valves, and the like. Having operating data in real-time is critical to allow control room operators to see and respond to changing conditions and keep the pipeline operating at peak safety and efficiency.

Operating data fit into three categories:

- *Parameter values.* These are numbers generated by instruments measuring pipeline data, such as pressures, temperatures, and flow rates.
- *Device status.* Rather than a numeric value, these signals give the state of being of a device. For example, a valve may be open, closed, or in transition. (*Note:* To some technical people, *status* refers instead to the quality of the data, good or bad. In this book, however, *status* means the state of being of the device.)
- *Other nonparametric data.* This includes alphanumeric labels for valves and other devices.

An instrument or device can have both status and value. The status of an orifice meter may be "in use" or "on," and the meter may register the value 100 Mcf per hour. To monitor and control a natural gas pipeline, the operator needs both the status and the value.

Points. The term *points* is used extensively when discussing pipeline control room operations. The origin and destination of data going to and coming from a device are called *points*. For example, the incoming motor-operated valve at a city gate in Berlin can have five different points, two command (open and close) and three status (open, closed, and in transition.) The valve itself is one point, with a *point tag* or a *point name*. All the data concerning this valve are transmitted from the *origin point* (the valve) to a *destination point*, the place in the computer where it is captured and stored in a *points database*.

The purpose of this discussion of points is to emphasize the care control system engineers must exercise to keep all this straight. Control systems have hundreds or even thousands of points. A points database or system administrator keeps track of all the points. Devices are listed on the points database and commonly shown on a process and instrumentation diagram (P&ID) for easier reference. During SCADA installation, technicians carefully check to ensure all points are listed and connected properly at each end. In the Berlin valve example, the point list is used to ensure the valve that shows up as the Berlin gate station incoming valve is really that particular valve. With carelessly installed crossed connections, the control room operator might open the wrong valve or no valve at all.

Following original construction, a points database is continually updated. If a new natural gas delivery point is added, the points must be correctly added to the points database and all devices tested and connected correctly. If any of these points become confused or supply bad data, information can be misinterpreted or incorrect pieces of equipment can be started and stopped. Because of the potential consequences of errors in the points database, access is strictly restricted.

Events. An important function of the control system is detecting, displaying, and recording *events*, which are changes to the pipeline system or actions taken by an operator or by the SCADA system.

Examples of changes to the pipeline include the following:

- Cold weather increases system demand.
- Transmission line maintenance reduces available pressure.
- A natural gas customer begins to withdraw an additional 10 Mcf per hour.
- A contractor ruptures a pipeline, causing a pressure drop.

Examples of operator-generated events occur when the operator generates commands to achieve the following:

- Adjust a control valve to raise pressure and increase flow.
- Open a valve to redirect flow.
- Place a remote device in maintenance mode.

Examples of system-generated events are the following:

- Files are downloaded to a backup device or location.
- A system device, like a printer or operator station, goes off-line or comes back online.

Events can be very interrelated. An operator event, such as adjusting pressure, results in system events, such as flow and pressure increasing or decreasing. Events may also trigger alarms. All this information is captured in a real-time database and is displayed to the control room operators on their screens. Then it is stored in a historical database for documentation, troubleshooting, and training.

Configuration tool

Any database needs to be configured for the specific application. The real-time database is no exception. Like a spreadsheet, the real-time database starts as a blank page, with rows and columns waiting for data. The users configure it (the spreadsheet or the real-time database) by writing instructions or commands in the case of the SCADA database, or equations in the case of the spreadsheet, using functions and macros. The functions and macros available to the programmer for writing instructions and commands are what make up the configuration tool. The easier and more intuitive the tools, the better users like the spreadsheet or SCADA database.

Developing, updating, and maintaining SCADA calls for, at minimum, a graphics editing program to build the screens graphics and a database configuration tool. Like a spreadsheet, the easier and more intuitive the tools provided by the vendor, the better the developer (and probably the SCADA user) will like it.

Application interface

Other applications use data stored in these databases. That calls for an interface or hook to allow data to go from the databases to the application. These hooks are analogous to the interface that allows customers to download their personal banking data from a bank's server to the money management software on their own computer. The customer does not see the hook but cares a great deal about whether or not the interface works well.

Like banking data, the SCADA interface has to be secure. Any time an application accesses the database, the interface must protect the database from outside intrusion.

Historic database

From the real-time operating database, some operating data is archived in a historical database. What are historic data used for? Changes happening over time may not be spotted using only real-time values but may become evident when trends are examined, leading to better understanding and optimization of the system. Logs documenting control room operator commands are also valuable tools for troubleshooting problems and understanding control room operator performance.

Over the last half century, historical databases have evolved from paper to mechanical charting to punch cards to magnetic tapes to disks to massive storage units. Accessibility increased at every stage.

Not every piece of data from the real-time database is transferred to the historical database. Again, the control philosophy dictates how much and what data are stored. Pressures may be read quite frequently but only changes (and when the changes occur) might be stored. Each piece of data included in the historic database is *time stamped*, showing the exact time it was generated, a critical piece of information when troubleshooting or diagnosing problems.

Historical databases provide a cornucopia of training information. With the right applications, this information can be played back, allowing a trainee to see exactly what a more experienced control room operator did in a particular situation and what happened as a result.

Amount of data to gather

Early pipeline dispatchers had only rudimentary real-time information about the pipeline they were operating. Was the equipment running? Was fluid coming out the other end? About how much was moved over a period of time? If less was coming out one end than going in the other, they found out why.

Modern pipelines operate with much more information. Some is vital; some falls into the "nice-to-know" category. More information can result in better operation, but presenting too much information to a control room operator can be confusing.

Update frequency

The SCADA system receives data from thousands of devices. The data travel for most of the trip along a common route (microwave, fiber optic strand, or a phone line). Some values can change rapidly and frequently, but *status* typically does not. SCADA manages all this data to ensure they are available on the frequency needed and do not become garbled in the process.

To handle the job of gathering all this data in an orderly fashion, an established protocol dictates how often and when the data are gathered. The frequency and manner falls into one of three categories:

- Polling (also called *scanning*)
- Report by exception
- Fixed interval reporting

During a *poll* or scan, the central system asks a device for data. *Scan tables* establish how frequently the devices are polled. Pressures may be reported quite often; meter readings from a natural gas delivery point may be collected only once per day. The scan tables refer back to the points database for data origin and destination.

For *report by exception*, a device reports only a change in value or status. If a pump turns off, it alerts the control system. But as long as it is running, it does not report its status. Some method of periodic integrity checking is usually implemented for report by exception to ensure data changes are not lost.

Fixed interval reporting is self-evident. A device reports its data on some regular basis. (Some say fixed interval reporting is simply another example of report by exception; if the set time has not elapsed, the device does not report.)

Update process

The control philosophy establishes how, and how often, data are received from each point. The update process acts as traffic control for the polling, exception reporting, and fixed interval reporting. If devices do not respond when scanned, or do not report at their appointed time, the update process initiates a diagnosis. If it cannot determine the problem, it feeds that information to the alarm process that either generates logs for future action or notifies the control room operator.

The update process continually performs checks of data validity by comparing selected incoming data to past data for the same device or to preset data.

Corrupted data

What happens when some of the data coming into the control system do not make sense in the overall context of the other data? Corrupted data is rare, but missing, incorrect, or old data can happen due to communication failures or failures in the field devices or network. The response can vary between two extremes when data appear to be corrupted. The control room operator can assume it is incorrect and simply ignore the suspect data, or the operator can immediately shut down the system until the problem is located and resolved.

The usual response is somewhere between these two options. Corrupted data may be an indication of an abnormal situation. SCADA systems contain validity checking routines that attempt to determine if data are good or bad. The bad data are excluded from the system and an alarm or warning is generated. Control room operators call out technicians to locate and troubleshoot the problem. In any event, control room operators cannot ignore seemingly bad data because they may in reality be accurate, and a problem may exist.

Indicators, alarms, and alarm filtering

Indicators display information that show what is happening on the pipeline. They can be lights, numbers, colored lines, animation, and sometimes audio cues (a bell ringing when an event is complete).

When indicators go beyond prescribed limits, they can become *alarms*, a visual or audio alert to a control room operator that an upset condition exists. Alarms are vital, but they can confuse and frustrate control room operators if they are not designed properly.

An alarm philosophy drives the alarm *filtering strategy*. Which conditions justify an audible alarm versus just an indication, perhaps one that blinks? Since upset conditions normally trigger many alarms, how should these be filtered so the control room operators get the information they need without overloading them with alarms?

Like the blinking seatbelt light in a car, frequent alarms can become invisible. Control room operators may not pay attention to an alarm when they should. Control room operators normally must physically acknowledge the alarm as a way to ensure they are aware of it.

The alarm process examines data as they come into the real-time database, comparing them to other pieces of data from the pipeline and to preset limits. If the alarm process detects a potential problem, it either creates an alarm immediately or pauses and double-checks the data validity from the next scan, to ensure before issuing an alarm that there was not a data glitch.

Alarms are prioritized by the alarm process. A visual alarm, such as a blinking light or a message on a screen, or an audible alarm, or both, may be issued. The process also records details about the alarm, such as when the alarm was created and when the control room operator acknowledged it (fig. 10–14).

Most alarm systems require the operator to acknowledge alarms by clicking a mouse or pressing a button. The alarm stays on until it is cleared. Some alarms can be cleared by the control room operator, while others require higher levels of supervision. Alarms can also come from applications like leak detection.

Fig. 10–14. SCADA alarm screen. The operator highlights the alarm or alert and then clicks on the button at the bottom of the screen to acknowledge the alarm. The system carefully monitors which operators acknowledged which alarms.

SCADA performance

Pipeline grids operate continuously, so their SCADA systems must be available at all times. Performance metrics indicate system reliability and areas needing attention. Metrics include the following:

- *Update time.* The time from a change of state of a field device until that change of state is displayed on the control console displays.
- *Scan time.* The time required to access or poll all devices.
- *Communications latency.* The transmission time for data to move from the remote device to the host system.
- *Screen update time.* The time from the request for a new screen until that screen is shown on the display. (This is a bit different from update time, since this metric is system dependent as it uses data already available in the system.)
- *Command processing time.* The time from when the HMI accepts a command until that command is processed or received by the field device.

- *Historical data access time.* The time from a request for historical data until it appears on the display.
- *System utilization.* The time the computing system resources are available at different SCADA load levels. (This is a measure of the surplus capacity to handle processing problems or large processing loads.)
- *System availability.* The time all components of the system are available for use. *Note:* This could be a percentage of the total time elapsed while the system is functioning as intended, that is, all components running and available with no redundancy called into play. It could also be the percentage of time the overall system is available for use even if some components or processes have failed. In this case, redundant components have been called into use or the failed process is not critical.

Data sharing

Who should be allowed to access the data? Few people outside the control room need real-time operating data, but many people are interested in recent historic data. Receipt and delivery data, schedules, forecasts, flow rates, and capacities are commercially important, and pipeline companies share this data both inside the pipeline organization and with customers. In some cases pipeline companies are precluded by law from sharing some of the information because it might benefit one customer at the expense of another.

Reports and logs

Reports and logs relate closely to data sharing. SCADA systems generate various reports automatically and distribute them electronically or in paper form within the pipeline company and to customers. Within the pipeline company, various groups can request that ad hoc reports be generated. The data-sharing philosophy greatly impacts the decisions around report generation and distribution.

Logs are a bit different. Logs capture events and commands and are an important part of the historic database for troubleshooting, optimization, and training. They are routinely generated by the SCADA system and normally distributed only to those involved with operations. However, when an accident occurs, logs are of great interest to all types of constituencies, both within the pipeline company and outside.

Security

Security, and as it relates to SCADA, *cybersecurity*, is of importance from a safety and a commercial perspective. Restricting system access involves firewalls, passwords, encryption, intruder detection, and secure protocols. Control rooms, computer rooms, compressor stations, and other company facilities are sometimes secured with live guards.

Security, like the other design decisions, must be carefully considered to ensure the solutions provided are appropriate for the type and magnitude of threat.

Data flows

Adding in the various components discussed in the previous sections results in a more complete SCADA schematic (fig. 10-15).

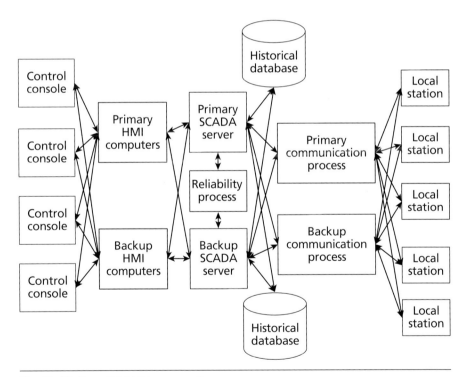

Fig. 10–15. Overall SCADA process flow

Figure 10–15 looks simple enough. Starting with the local station, data are acquired and transmitted to a central control room. From there control room operators send commands back to the local stations. Various computers and communications handle the data flowing both directions. Commands generated by the control room operator cause station devices to open, close, start, or stop. However, keeping all these instruments, computers, communications links, and other components properly calibrated and maintained is a huge task.

Summary

- SCADA systems are used to control many processes and are not unique to pipelines.
- SCADA standards exist but company preferences, facilities, and legacy operations mean each system is unique and custom designed.
- Some systems are complex, monitoring and controlling hundreds, even thousands of miles of pipe ranging from 24-inch or larger diameter high-pressure steel pipe to ½-inch diameter low-pressure house service lines.
- Other systems are quite limited and primarily monitor rather than control the pipeline grid.
- The most fundamental design decisions include human vs. machine decision making, manual vs. central decision making, the amount of automation, the amount of data to gather, how often to gather data, alarm filtering approaches, and communication and redundancy strategies.
- The human-machine interface, HMI, is the link between the control room operator and the control system, providing visualization and access to controls.
- Screens are the visual information link between control room operator and pipeline.
- Screens show three basic types of data: values, status, and information identifying the particular devices.
- Configuration tools are used to "build" screens.
- Real-time data are used to understand current operations, while historic data serve training, troubleshooting, optimization, and documentation purposes.

- Each piece of data is time stamped showing the exact time it was generated.
- Application interfaces are provided to make SCADA data accessible outside the control environment.
- Points databases are the road map to how data flow along the system and must be accurately maintained and updated.
- Performance measures provide insight into reliability and dependability.
- Protecting the SCADA system from outside intervention is a growing concern.
- SCADA, an acronym, will soon be accepted as a word, just like scuba.

chapter 11

Design and Engineering

I always thought the further along I got in engineering school the fewer assumptions I would have to make. Rather, I found the opposite is true.

—Ella Miesner, University of Texas, BS ME, 2008

The engineer waited in the conference room. Industrially framed pictures of city gate stations, gas storage facilities, district regulator sites, odorant injection facilities, and other related gas distribution facilities lined the walls, testaments to projects constructed over the past 40 years. Just recently the paper files, maps, and manuals held on tall shelves at the back of the room had been digitized, and the paper copies disposed of, although some engineers still had hard copies squirreled away in their desks.

The engineer had graduated five years ago from engineering school with a mechanical engineering degree, but until now, assignments meant optimizing and improving existing operations, not designing and building new facilities as she had hoped. Now she was about to become the project manager for a new city gate station. Thinking back over the past five years, she reflected on the fact that engineering and design are two different functions. Engineering is technical, and depends on equations and standards. Design requires operational knowledge and experience, which are not gained in engineering school, and is more of an art than a science. Now she realized the importance of understanding operations before moving on to the project manager role. She understood why the engineering manager had stopped by her office on her first day of work to offer the sage advice, "Codes and standards are not design manuals."

"Let's talk about your project," the engineering manager said, entering the conference room accompanied by the planning engineer. The planning engineer had taken the project, which was construction of a new city gate station for increased supply flexibility and reliability, from its conceptual stage through front end engineering design (FEED). The documents produced during FEED included a preliminary design, location, process flow diagram, schedule, list of long-lead-time items to be ordered soon, and a list of the remaining permits left to be obtained. The new project manager was looking forward to overseeing detailed design, detailed engineering, construction, and commissioning.

Now I finally get to build something, she thought as the meeting concluded. The new project manager left the conference room with a smile.

The smile did not always remain during the ensuing months as changing regulations, stubborn landowners, mounds of permits, inflexible local planning officials, late equipment delivery, contractors with financial problems, and inclement weather each took their toll. In the end though, the project was delivered on time and within budget, just as the heating season rolled around. As furnace pilot lights ignited the main furnace burners, oblivious homeowners went about their routines, and the project engineer smiled again.

Natural Gas Engineering Functions

Broadly speaking, natural gas pipeline engineering comprises four areas:

1. Optimization of day-to-day operations
2. Maintenance of existing lines and facilities
3. Modifications to existing lines and facilities
4. Design and construction of new lines, facilities, and control systems

Each of these functions must be performed with due regard for public and employee safety, supply reliability, gas quality, the environment, efficiency, and the regulatory regime. Items 1 and 2 are covered in their respective chapters. Accordingly, this chapter focuses on items 3 and 4. However, before moving on to pipeline and facility design and engineering, a short section dealing with the engineering and design process is in order.

Standards and Codes

Since the first street lighting with manufactured gas more than 200 years ago, engineers and operators have tempered the laws of physics with practical experience to produce engineering standards that are issued by industry organizations. With respect to standards, ASME International, a not-for-profit voluntary organization formed in 1880 as the American Society of Mechanical Engineers, defines *standards* as follows:

> *A standard can be defined as a set of technical definitions and guidelines, "how to" instructions for designers, manufacturers and users. Standards promote safety, reliability, productivity and efficiency in almost every industry that relies on engineering components or equipment. Standards can run from a few paragraphs to hundreds of pages, and are written by experts with knowledge and expertise in a particular field who sit on many committees.*[1]

When a standard is adopted by a governmental body, it assumes the force of law and achieves code status. The distinction between standards and codes is important and thus merits a bit more discussion. Standards are based on universal laws of physics. The engineering and technical factors included in the MAOP calculation (except for the units used), for example, are identical between nations and continents. However, the safety factor included in that calculation sometimes differs depending on the regulator or governing body having jurisdiction.

Standards often begin as "recommended practices," or suggestions rather than guidelines. Recommended practices morph into standards as more and more industry participants adopt the recommended practice, making its practice common rather than simply recommended. Some of the standards-issuing organizations include the following:

- American Gas Association (AGA)
- American National Standards Institute (ANSI)
- American Petroleum Institute (API)
- American Society for Testing and Materials (ASTM)
- ASME International (ASME)
- Standards Australia
- British Standards Institution (BSI)
- European Committee for Standardization (CEN)
- Gas Technology Institute (GTI)

- International Organization for Standardization (ISO)
- Manufacturers Standardization Society of the Valve and Fittings Industry, Inc. (MSS)
- NACE International (NACE)
- National Fire Protection Association (NFPA)
- Pipeline Research Council International (PRCI)
- Plastics Pipe Institute, Inc. (PPI)
- Russian standards (GOST, SNiP, OST, GTN, VNTP)

Providing a complete listing of standards is beyond the scope of this text. Nevertheless, one of the most widely used and longest existing standards in the gas distribution industry is ASME B31.8, "Gas Transmission and Distribution Piping Systems."

Many standards in addition to the previous list exist, and distribution system engineers use those that apply to guide their designs. Codes often incorporate standards by reference. The primary code governing natural gas distribution safety in the United States, for example, is contained in CFR Title 49, Part 192, "Transportation of Natural and Other Gas by Pipeline: Minimum Federal Safety Standards," which is enforced by the US Department of Transportation. In the United Kingdom, the regulations are contained in Statutory Instrument 1996 No. 825, "The Pipelines Safety Regulations of 1996," enforced by the Health and Safety Executive. Each country has its own (very similar) codes, which may include the same standards as those of another country.

It should be noted that standards and codes do not take the place of experience and operating knowledge. Engineers cannot just "take the standards off the shelf" and design an effective system.

The Engineering and Design Process

Whether designing the latest version of a smart phone, or a natural gas distribution system, the overall process includes the same basic steps as shown in figure 11–1.

The first three steps in the process, conceptual, feasibility, and FEED, enclosed in the dashed box in figure 11–1, are often quite iterative, with recycle and refinement back and forth. However, before the project transitions to the detailed engineering and design phase, the project should be well-defined with set design parameters. Project funding typically occurs at the end of FEED. Detailed engineering and design may overlap with construction, particularly in the areas of permitting and route selection. (In the smart phone analogy, manufacturing would replace construction.) As the construction phase winds down, the project is commissioned and then turned over to operations. Finally, at the end of its life, the asset is removed from service and retired or decommissioned.

Naturally, small projects such as the installation of a single service line do not have the same process rigor as larger projects, such as a new city gate station, for example. Regardless of the project, though, they all go through the same steps. However, detailed engineering and design may only take the few minutes required to download the previously designed meter and riser installation drawings and materials list from the company engineering library.

The process seeks to develop, design, construct, and operate the project with the lowest life-cycle cost, while keeping in mind public and employee safety, supply reliability, gas quality, the environment, and the regulatory regime. Figure 11–2 provides another perspective on the process and tasks involved.

Long linear features, such as roads, railroads, electric transmission lines, and pipelines face at least one unique challenge not encountered by many other industries—they have a lot of neighbors. The large number of homeowners and landowners, and the high level of interest in oil and gas pipelines, attract a lot of stakeholder interest in any pipelines, particularly new or large ones. Figure 11–3 depicts some of the stakeholders surrounding the pipeline company, shown in the center.

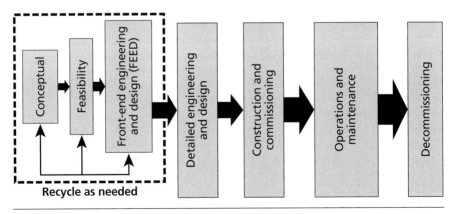

Fig. 11–1. Engineering and design process

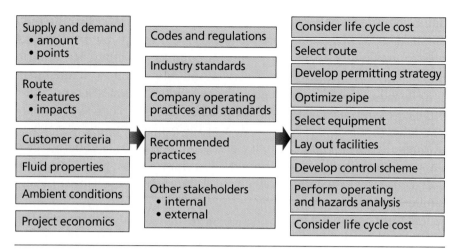

Fig. 11–2. Engineering and design considerations and tasks

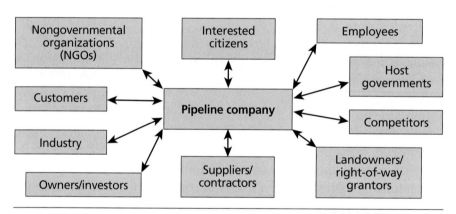

Fig. 11–3. Pipeline stakeholders

Each stakeholder or stakeholder group comes to the table with a unique agenda and different interests. It is up to the company to manage the process of sorting out and meeting the stakeholders' needs, as well as their wants.

Most local distribution service lines, and a large majority of distribution mains, operate within a designated and approved easement, providing the company the right to install and maintain the lines. Often mains are located under city streets or roads. Transmission lines typically must purchase their own easement or right-of-way (ROW), and the permitting process can be long and cumbersome. (Easements in some countries are called *way leaves*.)

Conceptual stage

Before a project is developed, it must be conceptualized. The concept may be as simple as a real estate developer with a new residential subdivision asking for natural gas service into the subdivision. Alternatively, a new supplier might approach the company's supply department, offering to connect to the system to supply gas to it. Or technology development might make smart meters or enhanced automation economically feasible. Regardless of its origin, each project is evaluated, formally or informally, to determine the project's feasibility, including the project risks and how to mitigate them.

Feasibility study

Some feasibility studies are quite simple and unsophisticated. For instance, a foreperson may have an idea that is within his approval authority. If he thinks it is a good idea, he is free to assign a crew to accomplish the task or project.

Other feasibility studies are quite long, consuming time and resources. In either case, the objective is deciding whether or not to continue the project to the next phase. In broad terms, projects must have an economic return or improve safety, security, the environment, or customer and public relations. In a centrally planned economy, it must have a public policy benefit. These factors are examined elsewhere in this text and thus are not addressed here. Focusing on capital projects, the next step in the process is FEED.

Front end engineering and design

FEED is often referred to as *front end loading* (*FEL*) and may be divided into several iterative steps. Most construction projects, pipeline or otherwise, use some sort of FEED or FEL process to narrow and refine the project prior to entering detailed engineering and design.

FEED seeks to reduce the uncertainty associated with a project by addressing key areas of concern:
- Design basis
- Capital costs
- Operating costs
- Ability to obtain permits
- Constructability
- Schedule

- Other factors that might negatively affect the project

Following is an example of the stages associated with one FEL process, along with the activities conducted in each stage:

- Stage 1
  - Business planning
  - Market studies
  - Conceptual design
  - Project economics
  - Risk analysis
- Stage 2
  - Review alternatives
  - Preliminary design
  - Preliminary estimates (± 25%)
  - Project economics
  - Risk analysis
- Stage 3
  - Class A design
  - Hazards analysis
  - Definitive estimate (± 10%)
  - Project economics
  - Risk analysis

FEED is an iterative process that hones the numbers with due consideration in each stage for the economics, risks, and public policy issues involved. Larger and more complex projects, such as replacing a SCADA system or building a major new transmission line, call for rigorous FEED processes. Smaller projects, such as installing a distribution line to a small subdivision, may not involve FEED at all.

As the FEED process draws to a close, the information developed is captured on various documents for transmission to the detailed design team. Some of the documents prepared for the design team include the following:

- Process flow diagram (PFD)
- Piping and instrumentation diagram (or drawing) (P&ID)
- Schedule
- Definitive (or class A) estimate (usually ± 10%)
- Major equipment list
- Control strategy (see SCADA chapter)

- Specifications for long-lead-time items
- Project execution plan
- Design basis document

Considering and narrowing down the alternatives and refining the project requirements during FEED permits the detailed design team to focus within a more narrow scope. Consequently, they do not have to spend as much money examining a number of additional alternatives. Various professional organizations can be accessed as additional resources for project planning and cost control. Two such organizations are the International Cost Engineering Council, http://www.icoste.org/, and the AACE International (formerly the Association for Advancement of Cost Engineering), http://www.aacei.org/.

Detailed engineering and design

The detailed engineering and design stage for a simple service line connection could be as straightforward as using the company's service line connection standard and heading to the storeroom to get the parts.

Alternately, it can involve many engineers, designers, surveyors, land acquisition agents, lawyers, public relations professionals, and others as needed. This team may work for many months to produce the myriad of drawings, specifications, purchase orders, legal documents, operating procedures guides, and other documents required to purchase equipment, secure permits, and develop plans and drawings that communicate to the construction contractors how to construct the pipeline or facility. The major tasks accomplished during detailed engineering and design include the following:

- Implement operating and control strategies
- Finalize route and acquire right-of-way or way leaves and land (primarily for transmission lines)
- Confirm pressures and flow rates
- Finalize pipe selection if not completed in FEED
- Valve locations
- Detailed design and layout of facilities
    - Receipt and delivery stations
    - Compressor stations
    - Other stations
    - Storage

- Design and execute permitting strategy
- Reconfirm economics
- Design quality control process
- Prepare operating and maintenance manuals

As the detailed engineering and design phase comes to a close, the construction phase gears up. Often various activities that began during design and engineering continue into construction. Examples include the following:

- Route selection
- Permitting
- Procurement
- Manual preparation
- Other tasks not completed

During a recent pipeline facilities design class taught by the author, he commented to the class, "Wouldn't it be great if all of the operations and maintenance manuals would be completed during detailed engineering and design and before construction commenced?" The class members vehemently stated that would not be possible, but reluctantly agreed having a large share of the manuals prepared before construction started was a worthy goal and a hallmark of a well-designed project.

Design and Engineering

The job of the local distribution company is to deliver natural gas to the right location at the right pressure and of the right quality, consistently, day in and day out. Everyone in the company focuses on that goal. Consequently, one of the chief duties of the engineering department is to design and supervise construction of systems that accomplish that goal. Pipeline design and facility design are discussed in this section, with a brief discussion of control system and electrical design.

Pipeline Design

Natural gas transmission pipelines are constructed almost exclusively of steel. Distribution mains may be steel or plastic. New service lines are typically plastic. Some legacy lines are copper or steel and may serve for many years to come. Legacy cast iron and wrought iron lines remaining in the system are systematically being phased out over time. The steel versus plastic decision is usually based on the amount of pressure the pipe must contain.

Pipeline configurations

Gas pipelines are configured either linearly for one-way feed or in a network (sometimes called a *grid*) configuration. Grid systems typically have a primary direction of flow but can also backflow as pressures, supply, and demand points dictate. Transmission pipelines are nearly all linear, moving gas in one direction from point A to point B, with delivery and receipt points along the way. Local distribution lines are normally interconnected networks of smaller diameter lines (fig. 11–4).

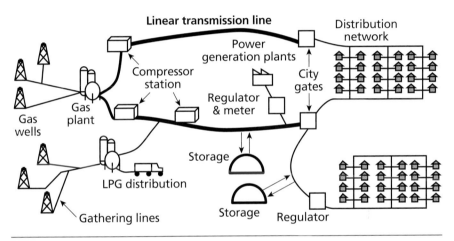

Fig. 11–4. Linear and network layout

Simplistically, the more supply connections, the greater the supply reliability, which is why natural gas distribution engineers always try to ensure each consumption point is supplied from at least two directions, which is known as a *back-fed* system.

Pressure

The Holy Grail of pipelines is pressure, or rather pressure differentials. All fluids, and natural gas is no exception, flow from higher pressure to lower pressure. As gas flows along, the molecules rub against the pipe wall and against each other, causing friction. Overcoming this friction takes work and uses energy, so the pressure in the system drops as the gas flows along. Delivery points out of the pipeline have lower pressure than upstream receipt points into the pipeline for transmission lines, and houses farther away from the city gate have less pressure than houses closer to the city gate for local distribution lines.

Hydraulic gradients. The pressure drop along the system, and the amount of pressure in the system at any point, are displayed graphically through the use of hydraulic gradients (fig. 11–5).

Figure 11–5 shows the pressure drop per unit (foot, mile, meter, or kilometer) dropping more quickly as the gas moves along from left to right. In other words, the pressure loss per unit increases as the gas flows along. Technically speaking, the slope of the drooping line in figure 11–5 is the pressure loss per unit, and the distance between the horizontal line and the drooping line is the pressure at that location along the pipeline if the pipeline elevation is flat.

Inquisitive readers might look at the graph in figure 11–5 and wonder if the shape of the line represents one of the shapes their high-school geometry teacher tried to drum into their heads. It is, as will be disclosed later. For clarity, natural gas engineers seldom draw gradients like those shown, but they repeatedly deal with the fact that gas velocity increases as the gas loses pressure in transit.

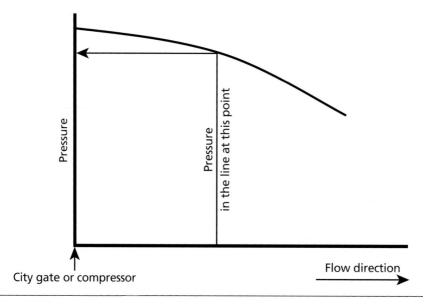

Fig. 11–5. Natural gas hydraulic gradient

Factors affecting pressure drop per unit of distance. While calculating pressure drop per mile is thankfully beyond the scope of this text, examining the factors engineers use when calculating friction loss helps explain important pipeline flow factors. It also leads to discovering the shape of the pressure loss curve. Factors include the following:

- Pipe diameter
- System
  - Pressure
  - Temperature
- Gas
  - Density
  - Viscosity
  - Compressibility factor
- Flowing velocity

Drinking from a smaller diameter straw is harder and slower than drinking from a larger diameter straw. Likewise, at the same flow rate, smaller diameter lines have more friction loss per unit of distance than larger diameter lines. At the same friction loss per unit of distance, smaller diameter lines have slower flow rates than do larger diameter lines. In other words, desired flow rates and pressures dictate pipe diameter. Figure 11-6 shows the relationship between diameter and pressure loss per unit.

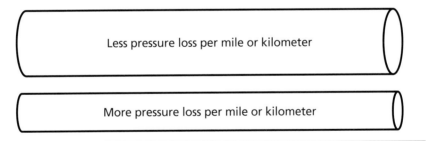

Fig. 11–6. Relationship between pipe diameter and pressure loss per mile or kilometer at the same flow rate

Relationship between temperature, pressure, and volume. Scientists of the 18th century with last names like Boyle, Charles, and Gay-Lussac discovered gases react to temperature and pressures in consistent ways. They developed equations to predict the relationship between temperature, pressure, and volume. Figure 11–7 demonstrates those relationships.

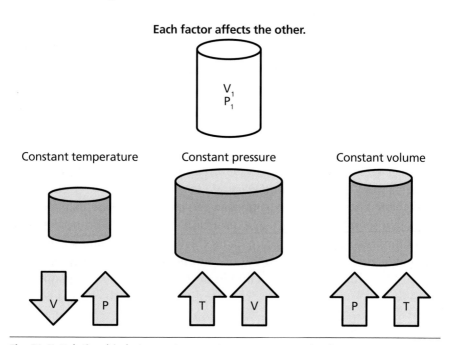

Fig. 11–7. Relationship between temperature, pressure, and volume

V_1 and P_1 represent the initial volume and pressure in a system at ambient temperature. Working from left to right:
- If temperature is held constant, as volume is decreased, pressure rises. Conversely, if pressure rises, volume decreases; that is, pressure and volume are inversely related.
- If pressure is held constant, as temperature rises, volume also rises, and as volume increases, temperature increases. In other words, temperature and volume are directly related.
- If volume is held constant, as pressure rises, so does temperature, and as temperature rises, so does pressure. So, pressure and temperature are directly related.

System velocity and pressure loss per distance. Engineers keep all these factors in mind, along with the gas density and viscosity, as they calculate system velocity and friction loss per unit of distance.

System flow velocity and friction loss per unit of distance are directly, but not linearly, related. As it turns out, friction changes by about the square of the velocity change. That is to say, when flow velocity doubles, friction loss per mile increases by a factor of four, and when flow velocity is reduced to half, friction loss per mile decreases by a factor of four. Finally, for those who like equations, $\Delta F = \Delta V^2$, where F is friction loss, V is velocity, and Δ means "change in."

Realizing that friction loss changes by the square of the velocity change, and remembering back to high-school geometry, the hydraulic gradient for gas is essentially a parabola. Recall from high-school geometry that $y = ax^2 + bx + c$.

The interesting thing about gas pipeline velocity is that each molecule in flowing gas pipelines is under slightly lower pressure than the molecule directly upstream from it, owing to the friction loss. As explained previously, lower pressure means the same number of molecules will occupy a greater volume. (*Quantity* is the number of molecules; *volume* is the space those molecules occupy.) Thus each molecule must be traveling slightly faster than the one immediately upstream. In other words, gas decompresses and the molecules accelerate as they move along. Accordingly, and obeying the laws of physics, pressure loss per unit increases as the gas flows along, resulting in the drooping hydraulic gradient.

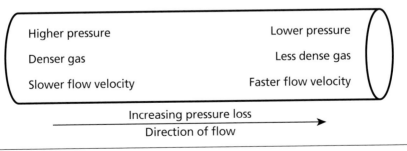

Fig. 11–8. Relationship between distance traveled, flow velocity, and pressure loss

Linear pipeline systems have one-direction flow, but flow direction may change in grid systems between seasons, from summer to winter, for example. Flow direction can even vary in parts of the system during the day. Consider figure 11–9, a simple system containing three city gates and four pressure regulators serving two subdivisions and an industrial plant. Depending on plant demand and the amount of gas coming in through each gate station, flow direction fluctuates in many of the lines, increasing the complexity of flow modeling. These changes in flows and flow direction mean local distribution engineers don't make use of hydraulic gradients on a normal basis—but they deal with the effects every day.

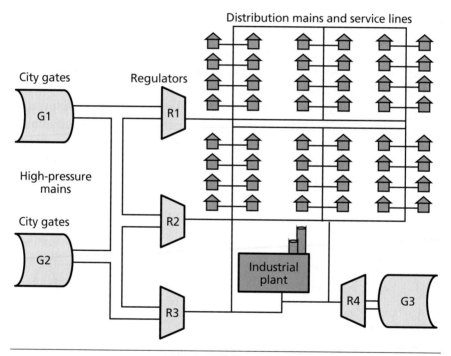

Fig. 11–9. Simplified service areas

Designers select pipes, regulators, and regulation points to ensure proper pressures on the warmest and coldest days of the year. To quote an experienced natural gas engineer, "There are three pressure levels, too high, too low, and just right. My job is designing systems where the pressure is 'just right' at each and every point along the line at each and every temperature, pressure, and flow rate the system encounters."

On a related note, increasing flow velocities have the potential to erode pipes and other components, so care must be taken to limit total pressure loss along a section that limits maximum flow velocity.

Pipe selection

Pipe is designed to contain pressure, and the amount of pressure pipe can hold depends on the pipe's characteristics:

- Strength
- Diameter
- Wall thickness

Pipe selection also depends on ambient conditions, such as the soil in which the pipe will be buried, in the case of underground pipe, or atmospheric conditions in the case of aboveground station piping. Temperature can affect both buried and aboveground pipe.

The laws of physics apply globally without respect to borders or the party in political power. Consequently, it seems safe operating pressures should be defined and calculated identically around the world. However, regulations vary between countries, and standards vary between organizations. Terms in use vary as well. The important concept is not vernacular or labels, but safety. Some of the pipeline pressure terms commonly heard, along with their acronyms, are the following:

- Maximum allowable operating pressure (MAOP)
- Maximum operating pressure (MOP)
- Safe operating limit (SOL)
- Maximum incidental pressure (MIP)
- Design pressure (DP)
- Operating pressure (OP)
- Peak level operating pressure (PLOP)

Defining all of these terms, and how they are determined, is beyond the scope of this text. Nevertheless, understanding how allowable pressures are calculated provides a useful perspective on the critical variables that influence safe pressure levels.

Steel pipe operating pressures. Using the term *maximum allowable operating pressure* (*MAOP*) as an example, in the case of steel, the MAOP calculation is

$$MAOP = (2t \times SMYS/OD) \times SF \qquad (11.1)$$

where
- $MAOP$ = maximum allowable operating pressure, expressed in pounds per square inch (psi) or kilopascals (kPa),
- t = wall thickness expressed in inches or millimeters,
- $SMYS$ = specified minimum yield stress of the steel expressed in pounds per square inch or kilopascals,
- OD = outside diameter of the pipe, expressed in inches or millimeters, and
- SF = a defined safety factor, usually expressed as a decimal.

Wall thickness, steel strength, and diameter, then, are the key variables influencing operating pressures.

Plastic pipe operating pressures. As a reminder, SDR is the ratio of the outside diameter of the pipe to its wall thickness, diameter and wall thickness being two key factors in calculating safe pressures. PE pipe with an SDR of 11 indicates the outside diameter is 11 times as large as the thickness of the pipe wall. The logical implication of this is that higher SDRs mean the pipe is better suited to lower pressures, since the wall is proportionally thinner.

The design pressure for PE pipe can be calculated according to the following formula:

$$P = (2S \times DF)/(SDR - 1) \qquad (11.2)$$

where
- P = design pressure (gauge) in psi or kPa,
- S = material strength based on its hydrostatic design basis as modified or limited by standards or regulations,
- DF = design factor established by standards or regulations, often either 0.12 or 0.4, with lower design factors signifying higher safety margins, and
- SDR = standard dimension ratio.

In actual practice designers look up on tables, rather than calculate, safe operating pressures.

Standards and codes may impose additional constraints on plastic pipe's operating pressure limits. Design engineers must be familiar with those constraints and design the pipeline or piping system accordingly.

Pressure/material considerations. Plastic pipe costs less that steel pipe, is less expensive to install, does not corrode (rust), and is more flexible than steel pipe. However, steel pipe, for the same wall thickness, can contain more pressure and has more structural integrity than plastic pipe. Engineers consider material characteristics, along with wall thickness, diameter, desired flow rates and pressures, and a myriad of other factors, as they design pipeline systems.

Isolation valves

Isolation valves are installed in transmission and distribution lines at selected points so segments can be isolated from each other for maintenance, construction, or emergency response. These isolation valves are commonly called *block valves* or *sectionalizing valves*. Minimum block valve spacing is established by codes and standards, but designers may include additional block valves depending on factors such as the following:

- Pipe diameter
- Operating pressure
- Location
- Population density
- Response time
- Number and size of branch connections
- Physical factors such as river, railroad, bridge, road, and stream crossings

Some block valves are automated for remote control. The acronym for remotely controlled valves is RCVs.

Another option is automatic shut-off valves (ASVs). As the name implies, these valves shut automatically under predetermined conditions, usually dropping pressure or increasing flow rate. Intuitively, smaller segments with RCVs or even ASVs result in faster response times and less gas released in the event of a leak. However, installing more valves is a trade-off as valves have more failure potential than just a straight piece of pipe. RCVs allow operator analyses prior to closing, but ASVs close automatically, meaning they might close when they should not. So when deciding block valve spacing and location and whether to install a manual valve, RCV, or

ASV, pipeline designers carefully evaluate the risks of installing additional valves versus the perceived safety those valves might provide.

Isolation valves are also installed at offtake and input points along transmission and distribution lines and at each service connection so individual service lines can be isolated from the distribution line.

Facility Design

Pipes transport the gas. Facilities do everything else. Designers select equipment, components, pipe, fittings, valves, instrumentation, and anything else needed to accomplish the particular functions required at each facility. Then they decide optimum placement for each of these items at the station site. Designers also select the control strategy to provide safe, reliable, efficient, and environmentally responsible operation of that station or facility.

Each facility, in concert with transmission lines, distribution mains, and service lines, must seamlessly and continuously supply the proper pressure (which also means the required amount) to each delivery meter on the day of highest demand. The highest residential demand day is usually the coldest day of the year. In the case of electrical power generation plants, the highest demand day is normally the hottest day. Pipeline and facility designers agree on the peak demands and then design to meet those demands. Hydraulic modeling software is available from various vendors to assist with hydraulic flow modeling. Before discussing facilities in detail, it is important to understand the repertoire of functions those facilities provide.

Facility functions

Some of the functions performed at natural gas transmission, storage, and distribution facilities include the following:

- Metering
  - Receipt
  - Delivery
- Compression
  - Origination
  - Booster
- Pressure regulation
- Odorant injection
- Storage

- Dehydration
- Liquid removal
- Temperature control

Natural gas facilities usually provide one or more of these functions. For example, a city gate station normally provides the following:

- Pressure regulation
- Receipt metering
- Temperature control
- Odorant injection
- Isolation

Facility designers start with the functions required at the station under design. Then they choose the components that provide those functions, and decide the position of each component on the plot of land. Next they decide how to connect the components together and how to control the entire station. All this information is placed on drawings and provided to the construction staff. While detailed design is well beyond the scope of this text, block diagrams of various facilities, along with additional items that should be considered in their design, follow.

Origination compressor stations

Most local distribution systems do not include compression. The gas arrives at the city gate with sufficient pressure already. Transmission systems normally include both origination and intermediate (booster) stations. Figure 11–10 is a block diagram of an origination station with four lines feeding the station.

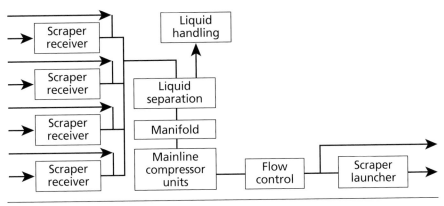

Fig. 11–10. Block diagram of an origination compressor station

As they design origination compressor stations, some of the additional items designers should consider include the following:

- Metering and calibration
- Treating and dehydration facilities
- Corrosion inhibitor injection
- Auxiliary equipment
- Cooling
- Fuel extraction
- Quality control

Figure 11–11 is a picture of an origination compressor station. In the building are three positive displacement compressors installed in parallel.

Fig. 11–11. Origination compressor station

Intermediate compressor station

As the gas moves along it loses pressure, so intermediate compressor stations are installed as needed to boost the pressure back to required levels. Figure 11–12 is a block diagram of an intermediate compressor station receiving from a looped line and compressing into a single outgoing line.

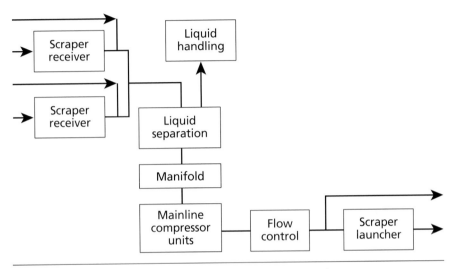

Fig. 11–12. Block diagram of an intermediate compressor station

As they design intermediate compressor stations, designers should also consider at least the following:
- Auxiliary equipment
- Metering and calibration
- Quality and testing
- Fuel extraction and quality

Figure 11–13 is a picture taken inside an intermediate compressor station containing two 7,500-horsepower centrifugal compressors installed in parallel.

Fig. 11–13. Intermediate compressor station
Courtesy: Williams Gas Pipelines; photo by Tom Miesner.

Natural gas interchange facility

Facilitating commerce and competition and optimizing system operations require connection points between natural gas transmission pipelines. These interchange facilities sometimes connect together two pipelines. At other times they connect multiple lines. Figure 11–14 is a block diagram of an interchange facility connecting two pipelines, either of which can deliver to or receive from the other, hence the arrows pointing both ways.

Fig. 11–14. Block diagram of a natural gas interchange facility

Figure 11–15 is an interchange facility. At times one line is the delivering pipeline, while at other times, it is the receiving pipeline. The metering is bidirectional at this facility.

Fig. 11–15. Natural gas interchange facility

Physical locations where multiple pipelines connect together are commonly called *hubs*. According to the US Energy Information Agency, as of this writing, there are 24 natural gas hubs in the United States. One of the best known and oldest is the Henry Hub in Henry, Louisiana. The longest operating physical transfer hub in Europe is the Zeebrugge Hub in Belgium. Figure 11–16 is the block diagram of a hub.

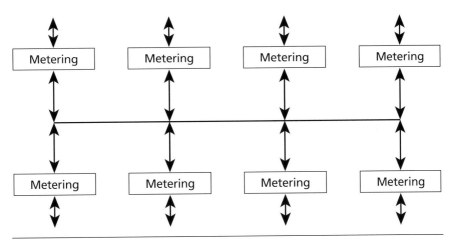

Fig. 11–16. Block diagram of a natural gas hub

Additional considerations at interchange facilities and hubs include at least the following:
- Flow conditioning
- Compression
- Liquid handling
- Pressure control
- Meter calibration
- Quality control

Natural gas storage

Natural gas is stored primarily underground in depleted gas reservoirs, caverns, or aquifers. Figure 11–17 is a block diagram of the aboveground facilities for underground storage. Figure 11–18 shows such a facility.

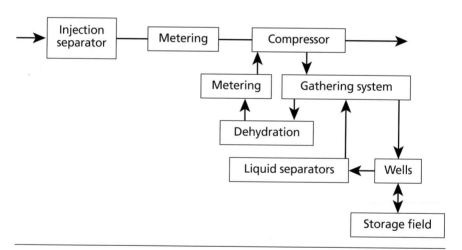

Fig. 11–17. Block diagram of aboveground facilities at natural gas underground storage facilities

Fig. 11–18. Aboveground facilities at a depleted gas reservoir currently used to store natural gas

Designers also consider at least the following as they design storage facilities:
- Wells may be injection, withdrawal, or both.
- The same compressors may be used for injection and withdrawal.
- The need for scraper launchers and receivers

City gate station

The first delivery point of natural gas into the local distribution system normally occurs at the city or town gate. Figure 11–19 is a block diagram of a city gate station.

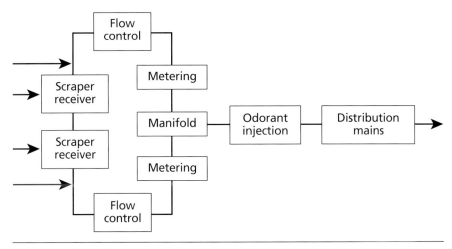

Fig. 11–19. Block diagram of a city gate

City gate design warrants at least the following additional considerations:
- Flow conditioning
- Meter calibration
- Gas quality testing and control
- Distribution main pressure control
- Gas temperature control

A city gate station is shown in figure 11–20.

Fig. 11–20. City gate station

Regulator stations

City gate stations discharge the gas into distribution mains. In general mains are divided into three categories, depending on their operating pressures:

- High-pressure mains (may be called *intermediate pressure* to differentiate from high-pressure transmission lines)
- Medium-pressure mains
- Low-pressure mains

Not every system has each main type, and the MAOP of each main classification may vary between systems, operators, and countries. Whatever the operating pressure, each step down in pressure requires pressure regulating stations. (In some locations, regulator stations are called *gas governors*.)

These are sometimes also called "district regulator stations" as a testament to past times when manufactured gas distribution operations were divided into districts for supervision reasons. Regulator stations normally have two pressure-reducing valves. One provides primary regulation, while the other serves as a backup, ready to take over if the primary regulator fails. Figure 11–21 shows a regulator station installed below ground level in a vault.

Several pressure reductions typically occur as the gas travels from the city gate to the final meter (fig. 11–22).

Fig, 11–21. District regulator station

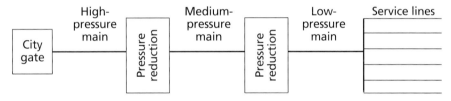

Note: not shown is the final pressure reduction at the final delivery point.

Fig. 11–22. Typical pressure reduction from city gate to final delivery point

Delivery stations

Local distribution systems serve four broad customer categories:
- Residential
- Commercial
- Industrial
- Electrical power generation

Gas usage varies across these groups, so the meter facilities vary as well. Installing new service to a small subdivision may not require any engineering at all, or at least no new engineering. In this case, installers simply check out standard parts kits from the storeroom and install each house meter following company-approved installation standards. Power generation plant delivery metering stations may be nearly as complex as a city gate. The complexity of metering stations for commercial customers is usually somewhere between that of residential customers and power plant stations.

Facility design guidelines

Standards and regulations guide facility engineering, but they do not much help with the design part of the task. Elegant design depends on knowledge and experience, making teaching design difficult. The following are from Vanderpool Pipeline Engineers Inc. Design Standards and are offered as an aid to those designing facilities.[2]

- Develop process flow diagrams (PFD) and piping and instrumentation drawings (P&ID) prior to beginning detailed design.
- Conduct thorough reviews of these documents with operations personnel early in the design process and involve an instrumentation technician with field experience as part of the design team.
- Develop operating procedures and cause-and-effect matrices to facilitate integrating safety and abnormal operations management into station design.
- Standardize equipment between stations to facilitate management of replacement and spare parts.
- Minimize turns and bends and maintain pipe, valve, and fitting diameters to minimize vibration within the piping system.
- Understand and install safety devices and emergency shutdowns to provide protection.
- Ensure sufficient clearance between pipe runs and equipment to allow easy access and maintenance.
- Think about step-overs and head bump dangers.
- Provide escape routes in case of emergency.
- Lay out auxiliary piping and electrical conduits to allow ease of access and safety.
- Consider an auxiliary bypass (small diameter) to facilitate maintenance and/or future station modification while continuing to allow the pipeline to flow, rather than requiring long lengths of blowdown.
- Install connections to allow system blowdown or evacuation.
- Provide supports as needed to ensure compressor flanges are not stressed and piping and component stresses are managed.

- Involve experienced operators in the design and review process to ensure operating needs are understood and met.
- Provide security as needed to prevent unauthorized access to the site.
- Design piping and supports to eliminate stress on compressor flanges.
- Complete a piping stress analysis to understand and provide for stresses.
- Be sensitive to the ambient temperature during fabrication and installation. Changes in ambient temperature cause great amounts of piping stress.
- Provide careful consideration for the amount of aboveground piping and underground piping.
- For safety and access, aboveground piping should be maximized, but turns, bends, and underground piping should be installed to manage piping stress.
- Provide lighting for security and maintenance purposes. Consider switch-controlled supplemental lighting for late-night call-out maintenance.
- Provide access to the control building just inside the gate and well separated from equipment in case of fire or other malfunction.
- Consider adding sufficient select fill to build stations higher than the 500-year flood level, as station flooding can be expensive and disruptive.
- Consider installation of remote cameras, along with "fire eyes" and gas detectors to provide visual checks from a remote control center.
- Provide station data recorders in the control building for all electrical, pressure, and temperature recording for efficiency diagnosis and other troubleshooting.
- Consider noise management of equipment for neighbors.
- Consider visual impacts and perhaps landscaping for noise and visual management.
- Evaluate electrical power requirements for immediate as well as future requirements.

Voltage available and design of incoming power can be significant limitations on future expansion and could affect station constraints for decades.

Facility Modifications

Designing facility modifications follows the same steps as designing new facilities, but designing on a greenfield site with nothing in the way is usually easier than designing on an existing brownfield site. With existing sites, the new components have to fit in and among existing components, adding constraints and increasing design complexity. In most cases, the existing facility must remain operational, adding more constraints and safety considerations.

Power System Design

Except for hazard classification, and perhaps the amount of redundancy, power design for local distribution facilities is quite similar to power design for most industrial applications. The power system normally consists of the following:

- One or more incoming transformers
- Backup generator
- Uninterruptible power supply (UPS)
- Low-voltage distribution
- High-voltage distribution

Figure 11–23 shows a typical low-voltage distribution system with an uninterruptible power supply.

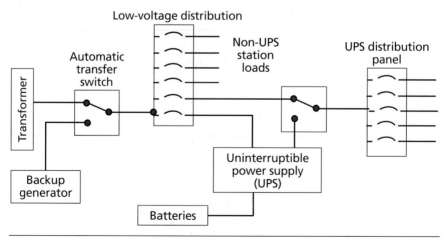

Fig. 11–23. Block diagram of a low-voltage distribution system for a local distribution facility

UPS systems typically supply power to critical instruments, controls, smart devices, communication equipment, and emergency lighting.

Control System Design

As with power distribution, local distribution control system design follows standard industrial design standards. It consists of designing how the hardware and smart devices are connected to best implement the control strategy decisions discussed in the SCADA chapter.

Other Engineering Tasks

In addition to designing pipelines, facilities, power distribution, and control systems, local distribution engineers frequently become involved with other tasks, such as the following:

- Optimization studies
- Hydraulic flow modeling
- System reinforcement
- Peak shaving
- Measurement
- Corrosion control

In addition, there are a myriad of other technical topics.

Summary

- Broadly speaking, natural gas pipeline engineering comprises four areas:
  - Optimization of day-to-day operations
  - Maintenance of existing pipelines and facilities
  - Modifications to existing pipelines and facilities
  - Design and construction of new pipelines, facilities, and control systems
- When a standard is adopted by a governmental body, it assumes the force of law.
- The engineering and design process seeks to develop, design, construct, and operate the project with the lowest life-cycle cost while keeping in mind public and employee safety, supply reliability, gas quality, the environment, and the regulatory regime.

- Pipelines attract a lot of stakeholder interest.
- A thorough FEED allows the detailed design team to focus and saves cost.
- Pressure, temperature, and volume are all related; when one changes, so do the others.
- There are three pressure levels: too low, too high, and just right.
- Facility designers start with the functions required at the station, choose the components and control system that provide those functions, and decide the position of each component on the plot of land and how the components are connected to control.
- It is important to develop operating procedures and cause-and-effect matrices to facilitate integrating safety and upset management into station design.
- Standards and codes do not take the place of experience and operating knowledge.

Notes

1. ASME, "About ASME Standards and Certification: About Codes and Standards," http://asme.org/Codes/About/FAQs/Codes_Standards.cfm.
2. David Vanderpool and Tom Miesner, Facility Design Considerations, Vanderpool Engineers, Inc., Engineering Standards, Denver, Colorado, 2011. Note: the design standards are proprietary and confidential and are used by special permission.

chapter 12

Construction

*Even if I knew that tomorrow the world would go to pieces,
I would still plant my apple tree.*

—Martin Luther

Reporting that the new town gate project was now flowing gas and accomplished under budget, the project manager expected hearty congratulations from the vice president of engineering. Instead, the VP asked, "What about the as-built drawings, operating procedure guides, inventories, inspection reports, completion reports, emergency response plans, and all the other documents required? Remember, the project is not finished until the documentation and procedure guides are complete." Two weeks later, the project manager returned to the VP's office to report all documentation and procedures were complete and operations training was in progress.

Documentation, although often considered boring and mundane, is the final step in the construction process and is critical to successful operations and maintenance. Another construction-related activity that is sometimes underappreciated is quality control, which is discussed toward the end of this chapter.

Introduction

The project flow chart from the previous chapter shows the construction phase immediately after detailed design and engineering. While each phase is included in a neat block with an arrow pointing to the next phase, in actuality, the dividing line between these phases is seldom clear-cut. In detailed design and engineering, the design is finalized, and in construction, the project is built, but spillover of tasks is commonplace, especially for large and complex projects. This overlapping of activities between phases happens most often for permitting, routing, right-of-way acquisition, and contracting. However, it can also happen with many other activities that begin in detailed design and engineering, or during even earlier steps, and carry through construction.

Sometimes the line between engineering and design and construction is clear, as in the case of small projects like residential meter sets. Residential meter sets are typically not designed individually. Rather, residential, and even some large industrial or commercial metering facilities, are routinely installed based on engineering performed many months, or even years ago, as reflected in company standards. For many construction projects, the design is so routine that company warehouses have prepackaged kits with the components required for the project.

Materials of construction are broadly either plastic or steel. Construction projects can be divided into the following three main types:

- Constructing plastic or steel lines and installing valves on land owned by others outside of fenced enclosures
- Building facilities, usually on land owned by the company, that are inside fenced enclosure
- Installing equipment and software to monitor and control the entire system

Each activity requires specific skills and procedures. Plastic line is lighter than steel and is generally fused together, whereas steel lines are normally welded together. Overarching each construction activity is the quality assurance (QA) or inspection process, which ensures each component is installed properly.

Contracting

Companies employ different construction contracting strategies. Some bid nearly all the construction work to outside construction companies, either on a project-by-project lump-sum basis or on a time-and-materials basis. Other local distribution companies maintain construction crews and supplement company crews with contract crews as needed.

Project-specific lump-sum bids require detailed bid documents containing drawing and specifications clearly delineating project scope and facilities design. Bid documents also contain the contract, and they list industry and company standards. Sometimes construction companies furnish a labor and equipment rate sheet for time-and-materials jobs, and they charge for the time worked and materials they purchase for the job rather than contracting on a lump-sum basis.

Permitting

Large natural gas transmission projects, particularly those crossing state, province, or international borders, usually require permits or certificates from one or more government's natural gas regulator. The US regulator is the Federal Energy Regulatory Commission (FERC), which issues or denies certificates following a lengthy review process. In Canada the regulator is the National Energy Board (NEB) for pipelines crossing provincial lines and provincial regulators, like the Alberta Energy Regulator (AER), for those that do not. Countries with centrally planned economies generally have considerably more streamlined permitting processes.

Permitting gas lines has a distinct environmental advantage over permitting oil lines. Any methane released from gas lines dissipates into the air, with no impact on surface or ground water, whereas oil line releases can contaminate surface and ground water, triggering extensive cleanup efforts.

Other permits

Beyond the certificate or permit from the federal regulator, an extensive list of other permits from a myriad of state, city, and local governmental agencies, all the way down to permission from the local planning commission for road crossings, are required. In some countries the permitting process involves public comments and extensive environmental, cultural, and endangered species reviews. Other countries employ less sophisticated processes. At the time of this writing, a flow chart detailing the FERC process is available at http://www.ferc.gov/resources/processes/gas-pipe-stor-perm.asp.

Acquiring rights-of-way

Transmission projects must normally acquire rights to install, operate, and maintain the pipelines from each landowner along the way. These rights are normally granted via a right-of-way (ROW) or way leave document. Companies pay a lump-sum fee giving them permission to install, operate, and maintain the line. In addition to the ROW fee, landowners are normally compensated for any damage to their property caused by initial installation or ongoing maintenance activities via a damages payment.

Easements

Distribution projects, such as installing mains and service lines along with supporting facilities in a subdivision, for example, are generally permitted at the time the subdivision is platted. The developer grants the distribution company easements as an inducement to install the lines. While there is typically no charge for the easement and construction-related damages, landowners in some cases receive compensation for damage caused during maintenance activities. As a consequence of these easements granted by the municipality or developer, permitting local distribution lines is normally less time-consuming than permitting transmission lines.

Unique challenges

Pipelines, and other linear assets such as railroads, power lines, freeways, and the like, are unique from facility permits. Linear assets often traverse thousands of different landowners and regulatory bodies, whereas siting of facilities like power plants and sewage treatment plants usually directly impact only a few landowners. Dealing with thousands of landowners and other interested parties means permitting and ROW acquisition can take years for large projects.

While it is true pipelines enable us to drive, stay warm, cook, and enjoy many other ubiquitous comforts and conveniences, individual landowners could receive all of comforts and conveniences if the pipeline went across their neighbor's property. An exception is the individual service line which, in nearly all cases, must cross the owner's lawn.

Satisfying landowners, regulators, and other interested parties often means the permitting process is drawn out and potentially litigious. Experienced pipeline project managers know the permitting process often requires significantly longer than actually constructing the line.

Plastic Lines and Connections

It seems simple enough: fasten lengths of steel or plastic together, dig a ditch, put the line in the ditch, include valves and takeoff points along the way, and then cover everything up. However, other buried utilities, aboveground facilities, streets, roads, inclement weather, traffic, and a host of other challenges combine to make each project unique, interesting, and often challenging.

Plastic mains

Mains start at the city gate. New distribution lines are typically constructed of plastic pipe, although in some limited instances steel may be used. Smaller diameter plastic pipe is delivered to the site wound up on a reel. Larger diameter pipe comes in straight lengths. Whether delivered to the location reeled up or in lengths, the ends must be connected together. *Fusing*, melting the ends of each pipe to be connected and pushing the ends together, bonding the melted plastic as it cools and solidifies, is often used to join lengths of plastic pipe. Compression fittings can also join pipe, although those are commonly used for repairs rather than new construction. When steel pipe is connected to plastic pipe, special transition pieces are used.

Various fusion techniques, including butt fusion, socket fusion, and electrofusion, are used; butt fusion is the most common. A butt fusion machine is used to square the ends of each pipe, and then a heating plate is inserted between the two ends, melting them. The heating plate is withdrawn and the ends are pushed together to form a tight bond (fig. 12–1).

Fig. 12–1. Butt fusion machine. Note the two blades on the circular trimming disk. Once the ends are trimmed so they precisely mate, a heating plate is inserted.

Fusion results in a small protruding bead of plastic (fig. 12–2).

Fig. 12–2. Fusion bead. The fusion bead is visible in the center of the picture between the two clamps holding the pipe in place.

Once fused together, the line is lowered into the ditch. Sometimes the pipe is the only underground utility in the ditch. To save cost and space, new subdivisions typically include the gas line along with electric, phone, and cable lines in a common ditch (fig. 12–3).

Fig. 12–3. Multiple utilities installed in the same ditch. Note the tracer wire on the spool and the warning tape installed above the gas line.

When utilities are laid in the same ditch, they are typically installed in a specific order and have minimum distances between them. As a word of caution, one should take the time to consult standards and local codes to clarify the order and distances. A quick look at social media for the answer to this question reveals confusing, ill-informed, and potentially dangerous advice.

Butt fusion works well when both ends of the pipe can be fit into the butt fusion machine. In cases where a line is connected to extend a buried line that remains in the ditch, electrofusion couplings are often used (fig. 12–4).

Fig 12-4. Electrofusion coupling. The black coupling has two connection points to which the leads from an electrofusion machine are connected. Internal elements installed in the coupling heat both the coupling and the pipe, fusing them together.

Smaller diameter pipe is often flexible enough to turn even tight corners. Large diameter pipe requires a thicker wall to achieve the same SDR as smaller diameter pipe. Consequently it is not as flexible and may require fabricated fittings to make the turn (fig. 12–5).

Fig. 12–5. Right turn made with a tee. This tee was connected by butt fusion. Note the fusion lines between the tee and the pipe.

Digging the trench through soft dirt is easy and is usually done with a backhoe. Excavation through rock requires special equipment (fig. 12-6).

Fig. 12–6. Rock saw

Rocks, or other debris in the ditch, can damage the pipe, so the ditch is often *padded* with sand. That is, sand is added to the bottom of the ditch before the pipe is lowered in. Then, more sand is added to cover the pipe prior to backfilling the ditch with the soils dug from the ditch.

Plastic branch connections

The same fusion techniques apply to installing branch connections or offtake points, but another important technique is added. *Saddle fusion* works like butt fusion, except that rather than melting the ends, saddle fusion melts the concave surface of the base of a saddle fitting. At the same time, a matching pattern is melted on the surface of the pipe. The two melted surfaces are held together and bond as the plastics cool (fig. 12–7).

Fig. 12–7. Preparing to install a saddle fusion tap. The concave surface of the tap, part of which is visible in the bottom of the figure, and the section of the pipe to which it will be fused, will both be heated to melt the plastic on their mating surfaces. Once the plastic is melted, the tap will be placed on the pipe and the clamp in the center of the figure will force the two parts together.

The tap in figure 12–7 serves a dual purpose. It connects the line and also contains a circular blade that is used to cut a hole in the line, allowing gas to flow through the connection.

Plastic service lines

The link between the distribution line tap and the meter—the *service line*— is installed using the same techniques as mains. Service lines are smaller in diameter than mains and are usually supplied in long lengths rolled onto a spool for easy transport. Thus the only connection points are usually at the tap and at the meter riser (fig. 12–8).

Fig. 12–8. Service line fused to the meter riser. From the riser, screwed steel fittings will connect the meter to the house.

Trenchless Construction Techniques

As pedestrians walk to work, the store, the theater, or wherever they are headed, they seldom consider the maze of utilities buried under the sidewalk or street, but natural gas and other utility workers do. Dealing with this maze, and installing new underground lines or structures, drove the development of so-called trenchless construction techniques, which do not require a full trench along the entire route.

Horizontal directional drilling

As lines traverse their route, they encounter obstacles that must be removed, worked around, or tunneled underneath. Horizontal directional drilling (HDD) techniques, pioneered in the oil field for guiding the drill bit to specific formations, made their way to the underground utility industry over the past 30 or so years. HDD rigs of various sizes are now common sights installing all manner of underground utilities.

First, a pilot hole is drilled from one side of the obstruction to the other. As the bit moves along, its position is tracked. The operator makes small directional adjustments so the bit exits at the desired location. As many sections of drill pipe as needed are screwed on as the bit moves forward. If the pilot hole is not large enough to pull the pipe through, a reamer is connected to the end of the drill string and pulled back to the drilling rig, enlarging the hole. Successively larger reamers are installed and pulled through until the hole is large enough, which is typically 1.5 times the diameter of what will be pulled through the hole. Figure 12–9 shows a typical HDD site. In this case the obstruction is a drainage ditch.

Fig. 12–9. HDD site. The pipe protruding from the ground in the foreground was installed into the previous hole. After the pipe has been pulled into the hole currently being drilled, the ends of each will be trimmed and connected together.

In many cases, the pilot hole is large enough so no reaming is required. Each HDD has unique aspects and must be planned carefully, including locating and avoiding other underground structures the drill may encounter.

Boring

Before the advent of HDD, boring was common technique. *Boring* involves digging holes on each side of the obstruction, lowering an engine with an auger attached into one hole, and essentially drilling under the obstacle to the other hole. Since the auger was not guided, old-timers tell stories of one or more large rocks in the path of the auger deflecting it towards the surface. This sometimes caused it to exit in the middle of the road rather than on the other side where it was first pointed. Boring and auger tracking techniques have improved with technology advances, and thus they now (usually) exit at the desired exit point.

Piercing

Rather than engines and augers, compressed air and piercing tools are sometimes used, especially for installing short lengths and small diameters. One arrangement involves a tool comprised of a piston within a casing. Compressed air moves the piston, and the impact of the piston drives the tool forward. Various refinements to piercing heads and alignment techniques have improved piercing tool accuracy to the point where piercing of up to 150 feet with acceptable accuracy is common. Some of the piercing tools can even be guided to improve their accuracy.

Service line installations using piercing tools commonly start at the meter location, where a small hole, called a *keyhole*, has been excavated. The tool is lowered into the hole and propelled towards the distribution line, where another keyhole awaits its arrival. When the piercing head arrives at the keyhole above the distribution line, the plastic service line is connected on and pulled back to the building.

Keyhole installations

Like laparoscopic surgery, keyhole construction involves gaining entry to the subsurface through a series of small holes. These holes are used to launch piercing tools, make connections to distribution lines, and conduct other operations. Keyholes are often *vacuum excavated* for safety, that is, the dirt is removed by a powerful trailer- or truck-mounted vacuum unit. The increased popularity of vacuum excavation means less direct contact with other underground structures and improved safety.

Continuing keyhole research funded by the industry and the public has resulted in the use of cameras and other specialized tools and promises to continue improving the safety and productivity of underground utilities, including natural gas distribution lines. At the same time, these tools minimize the impact to the public of construction and maintenance tasks.

Steel Lines

Steel pipe is supplied in lengths, generally about 40 feet, which are welded together along the ditch into a continuous line. Steel lines generally move gas at higher pressures than plastic lines, and typically do not have any low-pressure service lines connected to them. Along the way, however, steel transmission lines may deliver to power plants and other industrial facilities.

Steel welding, like plastic fusing, is an exacting task. The pipe ends are cut at a bevel to facilitate multiple weld passes. Prior to welding, the two pipe ends are lined up so the inside diameters of the walls are aligned and the spacing between the ends is precisely the right amount. Then they are clamped into place with either internal or external weld clamps (fig. 12–10).

Fig. 12–10. Two steel joints of pipe clamped and ready for welding. Note the external alignment clamp in the center of the figure holding the ends in proper position for welding.

After the welder makes the first pass, commonly called the *root* or *stringer* weld bead, the clamp is removed. Additional weld beads are added depending on wall thickness (fig. 12–11).

Fig. 12–11. Welder welding the cap or final bead on a steel gas transmission line

Lengths of steel pipe, commonly called *joints*, are welded together out of the ditch supported by wooden beams or skids to facilitate welding. While welding joints together out of the ditch is easier, it is not always feasible. As obstructions are crossed, and in congested situations, welding in the ditch may be the only alternative (fig. 12–12).

Fig. 12–12. Street construction. Note the patched street in the foreground extending into the background. The street is cut, and, one joint at a time, the pipes are welded together and lowered into the ditch.

After the joints of pipe are welded together, the coating immediately around the weld area is installed. Next the weld is tested, and the string of pipe is lowered into the ditch and then tied into the connecting lines. Finally, the ditch is backfilled and the surface is returned to its original condition. As with plastic pipe, the ditch may be padded to protect the pipe and coating from damage.

Facility Construction

While plastic is used for the majority of mains and service lines, steel is normally used to construct facilities, city gates, regulating stations, metering stations, and finally the delivery meter. Facilities have more components than lines, including meters, valves, flanges, strainers, odorant injections tanks, instruments, and potentially a myriad of other components. All these must be connected together to achieve the facilities function. These connections are made by welding, bolting, and screwing components together.

Facilities

The term *facility* is used to differentiate what is normally an aboveground installation from the pipe in the ground. Broadly speaking, natural gas local delivery systems consist of facilities, pipes, and a control scheme. Facilities can also be installed underground for convenience, safety, and noise control. One common facility is a metering and pressure reduction facility (fig. 12–13).

Fig. 12–13. Meter and pressure reduction station. This figure actually contains two facilities. The lighter color pipe and components on the right make up a meter station built by the transmission company. The darker assembly on the left is a pressure reduction station built by the LDC.

Engineers design facilities and prepare construction and equipment specifications. Designers produce detailed construction drawings. Procurement professionals purchase the materials and equipment.

Construction workers follow the drawings and specifications, welding together pipe, fittings, and flanges, thereby fabricating the necessary connections. The whole process is overseen by a project manager working for the LDC and a construction supervisor working for the contractor. In some cases the entire project is constructed by LDC employees without the use of contractors. Figure 12-14 shows fabrication work in progress.

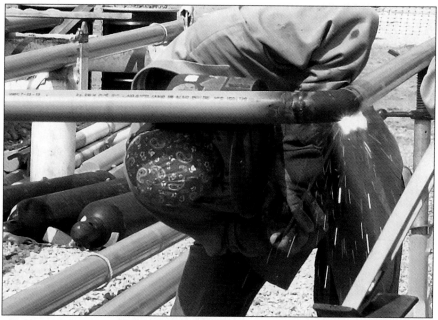

Fig. 12–14. Fabricating piping. Welding a "90" to make a right turn. The two pipes will be trimmed to the correct length and flanges will be welded to them. This fabricated piece will later be bolted into its assigned place. Note the adjustable stand supporting the pipe on the far right and level resting at an angle in the bottom right of the picture. Each pipe is carefully leveled and fitted together before it is welded.

Meter sets

Natural gas custody and ownership is generally transferred at the final meter, which is constructed by the distribution company. Smaller delivery meters, to individual houses for example, are commonly assembled by screwing fittings together and connecting pressure control and metering equipment (fig. 12-15).

This facility was likely installed by a LDC service person following a standard drawing that may have been produced several years ago. Many facilities are not custom designed.

Fig. 12–15. Fabricating piping. Two meters delivering to two different businesses. Note the third connection spot in the center of the picture and the ATM around the corner.

Quality Control

Engineers and designers take the first quality control (QC) or quality assurance (QA) steps. They work hard to ensure quality design, while purchasing professionals carefully specify pipe, equipment, components, and control schemes. In the same vein, construction workers inspect and test installations to verify that the quality approach begun during the design phase carries through to the finished product.

Qualification of personnel

One of the first QC steps is ensuring the person performing the work has been trained and is qualified to complete the task safely and at a high level of quality. Performing the task involves knowledge and skills; merely knowing how to complete the task is not enough. The person must transform that knowledge into the skill to complete the task. That is why, in addition to testing for knowledge, testing often involves an experienced trainer watching the worker physically performing the task. Each person working on the project must be certified for the tasks they will perform prior to starting the job. In many cases, industry standards specify how competence is proven and documented.

Inspection

Even qualified and motivated construction workers sometimes are tempted to take shortcuts, making on-site inspection during construction essential. LDCs assign their own employees or hire trained inspectors to physically watch the work as it progresses (fig. 12–16).

Fig. 12–16. Inspection. The inspector in the foreground observes excavation along a city street.

The inspector serves as the company representative on-site to make decisions when unexpected conditions are encountered.

When the situation encountered during construction is different than expected, inspectors are often called upon to make design change decisions. These decisions require understanding not only what the designer intended, but also how lines and facilities work and why natural gas behaves the way it does. Without a thorough understanding of operations, inspectors may approve changes that intuitively seem correct but unintentionally build hazards into the system. Companies employ a management of change (MOC) process to document changes to design, equipment, and operations. MOC processes require documentation of changes along with approval of those changes at the proper management level. In addition to inspecting and coordinating day-to-day decisions, inspectors also assist with collecting detailed construction-related data for incorporation into a data management system.

Testing

Beyond physical inspection during construction, various tests are performed to ensure construction quality.

Pressure testing. One common testing technique for distribution mains, service lines, and related facilities following construction is to pressure test them with air. The line is completely isolated and then pressurized to predetermined levels, such as 100 psi for mains. The pressure is monitored for specific periods of time to verify the line holds pressure (fig. 12–17).

Fig. 12–17. New distribution main under pressure test. Note the pressure gauge in the top background and the tap in the foreground connecting the main to the pressure gauge.

Steel lines are often tested *hydrostatically* after completion; that is, they are pressure tested with water rather than air. Small temperature changes affect pressure more during hydrostatic tests than during air tests, so hydrostatic testing requires careful monitoring and accounting for pressure changes caused by temperature changes. If not properly monitored, rising pressures caused by rising temperatures can mask very small leaks.

Pressure testing either with air or with water is rather straightforward if performed immediately following construction, as the new line and facilities are still isolated from the existing lines and facilities. Pressure testing conducted later becomes more complex and can cause serious service interruptions.

X-ray and ultrasonic testing. Steel line girth welds, that is the welds around the circumference of the pipe, are commonly tested using X-rays. This is the same technology as used for medical X-rays, but girth weld X-rays look for lack of fusion, lack of penetration, and contaminants in the weld or other defects that could affect pipe integrity. Film is fastened to the pipe and exposed to radiation for specific periods of time (fig. 12–18).

Fig. 12–18. X-raying a girth weld. The box in the center of the picture contains a radioactive element. The element is shielded in the box until the technician opens the box, exposing the film. A long cable extending from the back of the box allows the technician to stand a safe distance away.

Ultrasonic testing looks for the same defects, but uses ultrasonic technology.

Leak survey and testing. For many years soap solutions were used to test connections for leaks. The solution was applied to connections, and any escaping gas raised bubbles, indicating a leak. Even today, soap solution is still used in many cases. The various leak detection equipment discussed earlier in this text is also used extensively to ensure tight, leak-free connections.

Safety

Safety means everyone goes home at the end of the day in the same condition they arrived at the work site. Construction safety books and guidelines abound, so this text does not cover safety in detail, but only reminds the reader to always understand and follow safety procedures and use the proper personal protective equipment (PPE). Flame retardant clothing (FRC) is especially valuable as PPE; just ask anyone wearing it during a fire. As a final note, the leading cause of occupational fatalities is still vehicular accidents.

Data Collection

Data analysis leads to information, which leads in turn to knowledge that enables informed decisions. Broadly speaking, local distribution line data collected during the construction process consist of two categories:

- *Attribute data.* Information about pipe size, diameter, SDR, material of composition, date of manufacture, technician fusing the pipe, serial number of components, etc.
- *Geospatial data.* Location of the line and facilities, relationship to other facilities including roads, streams, and other utilities, landowners, and the like.

Of course data capture can encompass more areas than just the two above.

Over the past 20 or so years, as a consequence of improving geospatial capture and manipulation capability, integrity management and emergency response programs have grown considerably more robust. Now, mobile devices read asset data directly from bar codes or other sources and tag this data to location-specific coordinates. The data are then downloaded into corporate data systems, where they can be overlain with other information about the lines and facilities.

As-Builts and Inventories

Changes are often made during line or facility construction, so as-builts, the original construction drawings modified to reflect the changes, are produced during construction and completed shortly after construction is finished. As-builts are key tools for integrity management, emergency response, and line-locating purposes. Careful inventories are made of each component, piece of equipment, length of pipe, and fitting installed. This information is important for asset accounting purposes, including property tax and valuation purposes.

Operating Procedure Manuals

During detailed engineering and design, the engineers and designers, in collaboration with experienced operations personnel, began the process of developing operations procedure manuals for the lines and facilities under design. Often this means simply updating or adding to existing manuals. At other times it means writing completely different manuals. Whatever the case, these manuals are important training and operating reference aids, so careful preparation and periodic review are required to ensure they are always up-to-date.

Summary

- Engineering and design activities may overlap with construction activities, most often for permitting, routing, right-of-way acquisition, and contracting.
- Large natural gas transmission projects, particularly those crossing state, province, or international borders, usually require permits or certificates from the natural gas regulator or regulators of the countries involved.
- Permits from a myriad of state, city, and local governmental agencies, all the way down to permission from the local planning commission for road crossings, are usually required.
- The right to install mains and service lines is generally provided for at the time the development is platted.
- New distribution lines are typically constructed of plastic pipe, although in some limited instances, steel may be used.
- Various fusion techniques, including butt fusion, socket fusion, electrofusion, and saddle fusion, are used.
- Trenchless construction techniques (HDD, boring, piercing, and keyholing) that do not require a full trench along the entire route make construction and maintenance less costly and less intrusive.
- Lengths of steel pipe have their ends aligned and are then welded end to end.

- Steel is normally used to construct facilities, city gates, regulating stations, metering stations, and finally the delivery meter.
- Facilities have more components than lines, including meters, valves, flanges, strainers, odorant injections tanks, instruments, and potentially a myriad of other components.
- Inspection and testing are two important QA activities.
- Always understand and follow safety procedures and use the proper personal protective equipment (PPE).
- Data analysis leads to information, which leads in turn to knowledge that enables informed decisions.
- As-builts are key tools for integrity management, emergency response, and line-locating purposes.
- Manuals are important training and operating reference aids.

chapter **13**

Business Models and Expenditure Decisions

I have never known much good done by those who affected to trade for the public good.

—Adam Smith

Over the roughly 200-year history of the gas distribution industry, from manufactured gas to natural gas, the business model has continued to evolve. In nearly all cases the industry is highly regulated with respect to the allowable rates and the services it must provide.

Laws and regulations vary from country to country, and even within countries, so the LDC business model varies as well. This text does not set out the business model for any particular political or economic system but instead provides a framework for understanding the industry's business drivers and economic challenges as expenditure decisions are made by management.

Ownership

LDC ownership falls mainly into two categories: investors expecting a financial return, and governmental agencies that may or may not expect a return. In addition to the two main classes of ownership, some local distribution systems are *cooperatively owned*, that is, they are owned by those who are served by the system. In this case any excess funds generated may be returned to the owners. In other cases the distribution system may be a *master meter system*, defined as a natural gas pipeline system for distributing natural gas for resale within areas such as mobile home parks, housing projects, or apartment complexes, where the operator purchases metered gas from an outside source. The natural gas system supplies the ultimate consumer, who either purchases the gas directly through a meter on the property or by other means, such as service included as part of the rent.

Investors are either public or private, and governmental agencies include local cities or municipalities, and in some cases even countries. The owner may have its own operating employees. In some cases the owner contracts with a *distribution system operator* (*DSO*) that operates the system on behalf of the municipality. In France, for example, approximately 600 municipalities contract with DSOs to operate their natural gas distribution network in the framework of a public service agreement.[1] In Ireland, Ervia (formerly known as Bord Gáis Éireann) is a commercial semi-state company with responsibility for the delivery of gas and water infrastructure and services in Ireland.[2]

The DSO's mission is to guarantee access to the network, ensure natural gas flows through the network transparently and without discrimination, and manage the entity's economic viability. Of course just because a company is investor owned does not mean the government cannot be a majority owner of the company. For example, Gazprom is the major player in the Russian LDC marketplace. Each arrangement, whether investor ownership, municipal ownership, or some hybrid of the two, has unique advantages and disadvantages.

Bundled, Unbundled, and Open Access

Transportation of a commodity, in this case, methane, is the business of the LDC. In some cases the rate charged the customer for the gas and the delivery service is *bundled* or included together in one rate. In other cases the rate is *unbundled*, which signifies that the commodity price and transportation charge are listed separately. When the rate is unbundled, the LDC is often allowed to pass their gas purchase cost directly to the customer at actual cost. The transportation rate is then normally regulated. Another approach allows *open access*, that is, retail gas marketers sell gas to the customer, shipping it on the distribution system at the prevailing regulated rate.

As an interesting side note, since the transportation, and not the gas charge, is where LDCs make money, LDCs prefer low gas prices to stimulate demand through their system. High gas prices and decreased demand are normally detrimental to LDC earning. In the words of one

LDC executive, "We are not in the energy business; we are in the transportation and distribution business."

Revenues

Revenues are transportation rate multiplied by volume, plus other charges, in the case of bundled rate models. In the case of unbundled models, the commodity rate and transportation rate are determined separately. These two rates are then added together and multiplied by volume revenue. Finally, other charges are added to this revenue to arrive at total revenue (fig. 13–1).

$$\text{Revenue (\$)} = \left(\text{Transportation rate (\$)} + \text{Gas unit cost (\$)} \right) \times \text{Units} + \text{Other charges}$$

Fig. 13–1. Revenue calculation

However, as discussed in the previous section, the gas component of revenue may not earn any return, as the LDC simply pays for the gas and then receives a like amount from the consumer.

Volumes

Volumes are measured in standard cubic feet or standard cubic meters delivered to the location. In many cases the volumes are converted to energy content units such as dekatherms or kilowatt hours and charged accordingly. Volume or energy content is the first component of revenue calculations.

Rates

Granting an exclusive regional franchise to a distribution company is common practice. Only one provider means capital and operating efficiency—but creates market power. To avoid excessive charges by the exclusive provider, rates are regulated, normally using the cost of service (COS) approach. COS theory of rate making says a business should be allowed to recover its costs and earn a fair return on its invested capital. This seems quite straightforward, but each regulatory entity seems to interpret things just a little differently. Covering all the nuances between regulators across many countries is well beyond the scope of this text.

At a high level, though, COS establishes the allowed revenue and then allocates that revenue across all the units moved. The general steps for establishing required (or allowed) revenue are shown in figure 13-2.

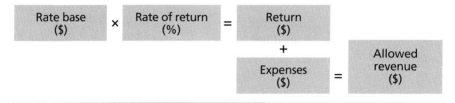

Fig. 13–2. Allowed revenue calculation

Each step is explained briefly in the next sections.

Rate base. Establishing the rate base on which to calculate the return involves the following equations:

$$\text{Gross plant} - \text{Accumulated depreciation} = \text{Net plant} \quad (13.1)$$

$$\text{Net plant} - \text{Accumulated deferred income taxes} + \text{Working capital} = \text{Rate base} \quad (13.2)$$

Weighted average rate of return. Three components enter into determining the overall rate of return:

- Capital structure
- Cost of debt
- Allowed rate of return on preferred and common equity

Capital structure is intended to reflect the debt/equity ratio of the entity owning the pipeline. Conceptually the cost of debt should be recovered (with no return) and a risk-adjusted return should be earned on the equity portion.

$$\text{Debt return}/\%\text{ Debt} + \text{Equity return}/\%\text{ Equity} = \text{Weighted average rate of return} \quad (13.3)$$

Return. The rate base is then multiplied by the weighted average rate of return as shown in equation (13.4):

$$\text{Rate base} \times \text{Weighted average rate of return} = \text{Return} \quad (13.4)$$

Allowed revenue. Expenses and taxes are then added to the return to determine allowed revenue:

$$\begin{array}{l} \text{Return + Operation and maintenance} \\ \text{expenses + Administrative and general} \\ \text{expenses + Depreciation expense +} \\ \text{Taxes (income and nonincome)} \end{array} = \text{Allowed revenue} \quad (13.5)$$

Operations and maintenance expenses. Accounting standards determine how expenses are classified (i.e., fixed or variable). For COS purposes, projected operations and maintenance expenses can be recovered if they are recurring, known, and measurable. Because operations and maintenance expenses can be associated with specific facilities, these costs can be directly assigned to pipeline services for cost allocation and rate design purposes.

Administrative and general expenses. Usually thought of as "overhead" expenses, these expenses may not be directly attributable to any particular facility or company. Rather they are incurred at the corporate level or by a subsidiary or head office group performing work on behalf of a group of pipelines. They are normally allocated among the various companies or facilities for which the work is performed

Depreciation expense. These expenses are considered a return of the investment over the useful life of the asset and are normally established by depreciation studies, which must then be approved by the regulator.

Taxes. Country and state or provincial income taxes on the equity portion of the return are calculated and added to the total cost of service. Other taxes, such as payroll and property taxes, are added to the total.

Rate per unit. Once the total allowed revenue is calculated, it must be distributed over all movements. If there was only one movement, the calculation would be as shown in equation (13.6):

$$\text{Revenue allowed/Units} = \text{Rate per unit} \quad (13.6)$$

These equations are only a guide. Individual regulators may take significantly different approaches and may allow (or require) separate calculations for overall system costs, individual customer costs, peak days, load factors, and the like as the rate is divided among movements and shipper classes.

Purchased gas

LDC gas supply departments purchase gas from a variety of sources under a variety of different arrangements and then transport it to the city gate under different transportation service agreements. All these transactions must be tracked in order to support the cost of the gas that is passed on to the customer. Many gas bills contain a purchased gas entry that is added to the transportation entry.

Other charges

A myriad of other charges may be added on for a variety of reasons, including administrative costs, taxes, and integrity surcharges.

Expenditure Decisions

Accountants divide expenditures into expense and capital for accounting purposes. In general, expenses are charged directly against revenues in the current year, while capital expenditures are depreciated over time.

Operators are mindful of these two categories, but divide expenditures a little differently. Operators focus on whether the expenditure is discretionary or not, rather than on whether it is capital or maintenance.

Nondiscretionary expenditures

These expenditures include items such as fuel and power, operating and maintenance people, small maintenance, and integrity expenditures, all of which are required in the short-term simply to keep the system operating. By definition, *nondiscretionary expenditures* cannot be delayed and must be spent. Accordingly, whether or not to make those expenditures does not require management to make a decision; the expenditures must be made or the system's ability to deliver will be impacted.

Discretionary expenditures

Management must decide if, and when, discretionary expenditures are made. Some discretionary expenditures yield an economic return, while others do not. Figure 13-3 differentiates between the accounting and operating view of expenditures and divides discretionary expenditures between those that earn a return and those that do not.

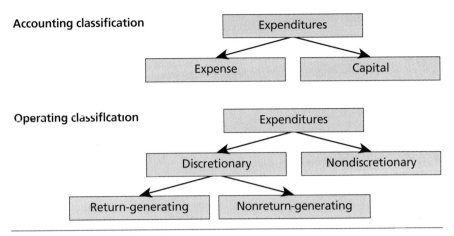

Fig. 13–3. Expenditure categories

Some investments fall into an "in-between" category. That is, they may generate a small amount of incremental revenue or earnings but not enough to make them economically viable on a stand-alone basis.

Return generating

The line between nonreturn expenditures and those that earn a *return*, or bring in more money than they cost, is often hazy. For example, does a marketing expenditure generate more business or simply maintain existing business? Rather than put a fine line on the question of whether or not any individual expenditure earns a return, for purposes of this text, the focus is on how companies consider discrete projects. The process is essentially the same for LDCs as for any business.

Expected value. Expenditure decisions are forward looking and carry a degree of uncertainty, giving rise to the concept of expected value, which is essentially the same concept as risk. Just as risk depends on consequences and probability, so expected value depends on the size of the anticipated revenue stream and the likelihood of that revenue stream materializing. The expected value of a $1 revenue stream that is certain to happen is $1. Similarly, the expected value of a $10 revenue stream that has a 10% change of occurring is $1. Expected value then depends on anticipated revenue and probability.

$$\text{Expected value} = \text{Function (Revenue, Probability)} \qquad (13.7)$$

Expenditures for projects that are less certain require larger anticipated revenue streams for the same expenditure than those with greater certainty. On a personal investing level, savings accounts, which are more secure, pay lower interest rates than do junk bonds, which might not pay back even the principal invested.

The size of the expenditure to achieve the revenue stream, not just the certainty and size of the revenue stream, is uncertain. Also, the further into the future the projection is, the less certain it is. Investment decisions then focus on developing reliable expenditure and revenue estimates and are often developed in ranges rather than points.

Rate of return. Put an economist, an accountant, and an investment banker in a room and ask them to define rate of return, and they would come up with at least three different answers. For purposes of this text, higher rates of return are better than lower rates of return, assuming both have the same likelihood of happening.

Project analysis. Analysis techniques vary between companies and across countries. However, in general, project analysis involves developing estimates for revenue and expenditures by year, which are then used to calculate cash outflows and inflows. One technique considers the internal rate of return (IRR) of projects. Table 13–1 contains a simple example.

Discretionary return-generating projects are approved (or not) by management after due consideration of investment, revenue, expenses, risks, rate of return, and potentially other factors.

Table 13–1. Simple IRR calculation

	Investment		Years			
		1	2	3	4	5
Revenue		€ 125,000	€ 128,750	€ 132,613	€ 136,591	€ 140,689
Cash expenses	−	€ 82,500	€ 84,975	€ 87,524	€ 90,150	€ 92,854
Noncash Charges	−	€ 15,000	€ 15,450	€ 15,914	€ 16,391	€ 16,883
Pre Tax Income		€ 27,500	€ 28,325	€ 29,175	€ 30,050	€ 30,951
Tax	−	€ 8,250	€ 8,498	€ 8,752	€ 9,015	€ 9,285
Income After Tax		€ 19,250	€ 19,828	€ 20,422	€ 21,035	€ 21,666
Noncash Charges	+	€ 15,000	€ 15,450	€ 15,914	€ 16,391	€ 16,883
Cash Flow	−€ 115,000	€ 34,250	€ 35,278	€ 36,336	€ 37,426	€ 38,549
IRR	17%					

Nonreturn generating

Discretionary nonreturn-generating expenditures are those which, delayed in the short-term, are unlikely to result in an immediate operating challenge or system deterioration. If delayed, however, mid-term or long-term, there is an increased risk of system failure that is difficult to quantify. Consequently, these expenditures present unique decision-making challenges. As shown in figure 13-4, nonreturn category expenditures can divided into those required to achieve the following:

1. Maintain long-term production capability
2. Comply with safety and other rules and regulations
3. Maintain risk at acceptable levels

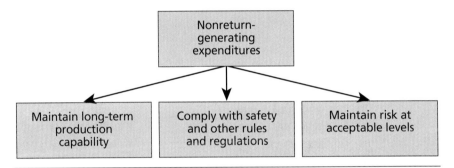

Fig. 13–4. Nonreturn expenditures

Item 1, the cost to maintain production capability, that is keep the system operating at its current capacity, is relatively easy to quantify. These are very close to nondiscretionary expenditures, with the consideration being one of timing. The prime decision-making criterion with these expenditures is whether or not the current budget can accommodate the expenditure. If not, they are usually delayed until they become nondiscretionary or until the annual budget has more flexibility.

Item 2, the cost to comply with rules and regulations can vary widely. In most cases, the regulatory requirements are reasonable, and companies would make those expenditures even in the absence of regulations. In some cases, there are differences of interpretation between operators and regulators, and even different regulatory agencies, which cause expenditure uncertainty. In many cases the extra cost to comply often comes down to the cost of the additional record keeping required to prove compliance. These expenditure decisions often are made either as a result of negotiations with, or orders from, the regulator.

Item 3 is maintaining risk at acceptable levels. At a high level, risk is rather easily defined as the combination of likelihood and probability. However, risk tolerance varies widely among people, and defining what the term *acceptable risk* means is difficult. Hardly anything is risk free. Managers use available tools to consider risk and attempt to prioritize expenditures to those that reduce the risk the most for the dollar spent.

Valuation

The five most common valuation methods for pipelines are as follows:
- Economic value
- Comparable sales
- Highest and best use
- Reconstruction cost new, or depreciated reconstruction cost new
- Book value

Economic value. The various ways of calculating economic value comprise the criteria used most often to establish the seller's bottom acceptable price and the purchaser's maximum purchase price.

Existing pipelines have a history of returns. Potential purchasers may base the price they are willing to pay on forecasts of volumes, rates, and expenses based on history. The forecasts are used to develop a future stream of cash flows that are then discounted to arrive at a value, called *net present value* (*NPV*). The discount factors represent the time value of money. Receiving $1 today is worth more than getting $1 five years from now, simply because of interest rates. Discounting future streams of cash flows by a factor equivalent to interest rates allows comparison of present and future values.

The promise of a dollar in the future also has more risk than getting a dollar today. That gets reflected in the returns from various kinds of investments. A secure government bond may yield 4.0%, while a bond issued by a pipeline company might deliver 6.25%. Pipeline companies as a group have a less favorable track record than the US government. A new pipeline into a new area has more risk than a line currently in operation. One has a track record and the other does not. Investors will demand that pipelines return a higher return than government bonds to reflect the higher risk, especially new pipelines. Thus, some companies will discount future cash flows with a factor that includes both interest rates and a risk factor to determine if the sum total of those discounted flows is more than their cash outlay as an investment.

Another economic valuation uses a multiple of the current cash flow, earnings before interest, taxes, depreciation, and amortization (EBITDA), or some other financial term. In this case, the agreed annual number is multiplied by a factor to arrive at the valuation. It may seem easy, but getting agreement between buyer and seller on the factor is a matter of intense negotiation. The choice implicitly includes both the rate of return each party expects plus the risk associated with future cash flows. The higher the risk (in the eye of the buyer), the lower the multiple should be. Just as in justifying an expansion or a new line, the assumptions, not the mechanics of calculations, are of critical importance.

In almost all cases, the buyer and seller make different assumptions. The seller may have a strategic imperative to dispose of the asset. The buyer may dream of expansion possibilities or volumes that the seller does not.

The calculations may seem simple, but the buyer and seller both go through laborious and time-consuming processes to decide what financial concept to use and which discount factors or multiples to apply.

Potential purchasers, particularly those operated as investment vehicles such as MLPs, also run cases with the new line consolidated into the balance of their existing portfolio to ensure ownership is accretive to earnings. Accretion is critical to increase distributions.

Comparable sales. Establishing value based on sales of similar items works well where similar assets, like houses, are sold every year, establishing a market value. It does not work well for pipelines because so many adjustments need to be made to arrive at an appropriate value, based on factors such as age, throughput, and the supply/demand outlook for the commodities being transported. Buyers and sellers may use similar sales as indicators to help them understand the range of values, but normally this valuation method only complements the economic value method in pipeline sales.

Highest and best use. This valuation technique is not normally used for pipelines. It is based on the notion that the current use of an asset may not be the best economic use. A group of houses may be bought for more than comparable houses and demolished to make way for a shopping center. In this case, the houses were bought to secure the land. The highest and best use for an 8-inch crude oil line with declining volumes may be to replace it with a 20-inch gas line. The 8-inch line might be purchased for more than its economic value as a way to secure the right-of-way for the 20-inch line. Highest and best use can be considered in combination with economic value for pipelines.

Reconstruction cost new. *Reconstruction cost new* (also called *replacement cost*) is the cost of rebuilding the same pipeline. It is valuable primarily as a check by purchasers to ensure they do not pay more than it would cost them to build the same or similar pipeline. Reconstruction cost new can also be used as a starting point from which to devalue an asset to account for its current condition. Depreciated reconstruction cost new can be a valuable tool during the negotiation process to help the buyer drive the price lower than economic value. If sellers expect to receive reconstruction cost new for a line, they may be disappointed if it exceeds the economic value.

There is one caveat with respect to reconstruction cost new as a ceiling. The fact that a pipeline already exists and does not need to go through the permitting process can be of great value. In some cases, it can drive the price above the cost to build a new line.

Book value. Book value is used by sellers to allow them to understand if they need to record a financial gain or loss for the sale. It is not particularly useful for purchasers, since book value has little relevance to what the asset is currently worth. Sometimes it is used to reallocate ownership between joint venture partners if the venture was set up to provide for this approach.

After all the valuation work is completed, price comes down to negotiation, and buyers always want to pay less, and sellers always want to receive more. In the seller's ideal world, if more than one buyer is interested in purchasing the pipeline, the seller can leverage them to achieve the highest potential price. The reverse is seldom true. In the unique market for pipeline assets, more than one pipeline in an area is rarely on the block simultaneously.

Summary

- Natural gas distribution is highly regulated with respect to allowable rates and services.
- Laws and regulations vary from country to country, and even within countries, so the LDC business model varies as well.
- LDC ownership falls mainly into two categories: investors expecting a financial return, and governmental agencies that may or may not expect a return.
- Rates are either bundled or unbundled.

- High gas prices and decreased demand are normally detrimental to LDC earnings.
- Granting an exclusive regional franchise to a distribution company is common practice.
- COS theory of rate making says a business should be allowed to recover its costs and earn a fair return on its invested capital.
- Many gas bills contain a purchased gas entry that is added to the transportation entry.
- Operators consider expenditures as discretionary or nondiscretionary.
- Project analysis involves developing estimates for revenue and expenditures by year, which are then used to calculate cash outflows and inflows.
- Risk tolerance between people can vary widely, and defining what the term *acceptable risk* means is difficult.
- Managers use available tools to consider risk and attempt to prioritize expenditures to those that reduce the risk the most for the dollar spent.
- The most common valuation methods for pipelines are economic value and reconstruction cost new or depreciated reconstruction cost new.
- After all the valuation work is completed, the price comes down to negotiation. Buyers always want to pay less, and sellers always want to receive more.

Notes

1. *Gas in Focus*, "Natural Gas Distribution System Operators in France," http://www.gasinfocus.com/en/indicator/natural-gas-distribution-system-operators-in-france/.
2. "Ervia Explained," Ervia, http://www.ervia.ie/who-we-are/ervia-explained.

chapter 14

Challenges for the Future

Neither a wise man nor a brave man lies down on the tracks of history to wait for the train of the future to run over him.

—Dwight D. Eisenhower

"Our job is safely, reliably, and efficiently providing each and every customer the natural gas they need every minute of every day, in an environmentally responsible manner," the company president told the news reporter during an interview brought about as a result of a supply disruption during a frigid day. "Of course, we have to do this in a manner which balances the demands from our myriad of stakeholders," the executive added. He then went on to say, "The expectation is perfection, and any supply disruption, safety hazard, or environmental incident, no matter the reason, is a failure."

Most people, when they hear *stakeholder* as applied to the natural gas distribution industry, immediately think of a customer. In the EU-28, according to Eurogas, there are a little over 120,000,000 natural gas customers.[1] The US Energy Information Administration estimated that in 2013 there were nearly 67,000,000 residential, more than 5,000,000 commercial, and almost 200,000 industrial natural gas customers in the United States.[2] The lesser developed parts of the world have fewer customers per capita but would still add significantly to the world total. That is a lot of customers to satisfy. But add to the stakeholder list city, state or provincial, and federal governments, and the balancing act becomes even more delicate (fig. 14–1).

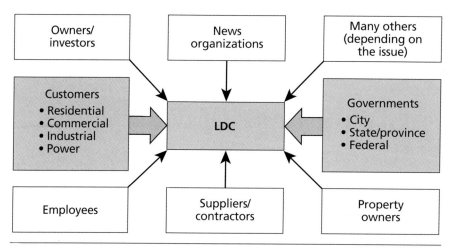

Fig. 14–1. Stakeholders

As one natural gas expert said, "Local distribution companies have millions of neighbors, and one of their biggest challenges is keeping all of them happy." The overall challenge, then, is providing safe, reliable, efficient natural gas delivery with due consideration for the environment, while keeping stakeholder needs balanced. This text discusses safety, reliability, efficiency, and the environment. It finishes with the overarching challenge of employee knowledge in the context of an aging work force.

Public Safety

Natural gas, in the proper concentration with oxygen, easily ignites and burns. This makes natural gas an ideal fuel but also poses two of its top safety challenges: preventing releases from occurring and preventing ignition when they do occur.

Excavation damage

One of the leading causes of hazardous natural gas pipeline releases in North America, Europe, and likely the balance of the world is excavation damage. Some party digging near the pipe hits it, damaging the line. The line strike sometimes punctures the line, causing a natural gas release. Excavation damage is not peculiar to the natural gas distribution industry. Phone lines, cable lines, fiber optic cables, water lines, and sewer lines are all subject to excavation damages. The industry is responding to the excavation damage challenge in several ways.

One Call. Going back at least as far as the mid-1970s, cities, states, and provinces began implementing One Call programs. These programs provided excavators the opportunity to call one centralized number several days prior to beginning work. The work location information was sent to utility owners, who checked their records and dispatched service personnel to locate and mark the utilities as needed.

Over time, these programs became more centralized, making them easier to use and publicize. In many cases, in the United States, these programs are now organized on a state-by-state basis. The One Call program in the state of Illinois, for example, is called JULIE, an acronym for Joint Utility Locating Information for Excavators. The state of Illinois provides a useful example, as JULIE is just for excavation outside the city of Chicago. Inside the city of Chicago, the program is called Digger—Chicago Utility Alert Network. Other political entities can have similar peculiarities. To avoid confusion and make excavation safer, the United States established 811 as a centralized, nationwide number.

Excavators can call 811 from anywhere in the United States a few days prior to digging, and the call is routed to the local One Call center serving that area. The excavator tells the operator where they are planning to dig and what type of work they will be doing. The affected local utilities companies are notified of the intent to dig and send a locater to mark the approximate location of underground lines, pipes, and cables, so the contractor knows what is below the surface and can dig safely.[3] Since utilities cannot always respond immediately, excavators should call several days prior to the time they want to begin work.

Similar location and marking schemes exist in other countries and cities to varying degrees of sophistication and centralization.

Excavator and utility consortiums. In 2000, to address the problem of underground utility damage in the United States, a consortium of companies banded together to form Common Ground Alliance (CGA). The specific aim of CGA is to reduce damage to underground infrastructure. A diverse group, the CGA includes excavators, locaters, and road builders, along with electric, telecommunications, oil, gas distribution, gas transmission, and railroad companies. CGA members also include One Call, public works, equipment manufacturers, state regulators, insurance companies, emergency services, and engineering and design professionals. CGA releases a new edition of *Best Practices* every spring, with updates that reflect changes in damage prevention in the light of new technologies.[4]

Cooperative excavation damage prevention is not unique to North America. Natural gas local distribution companies around the world strive to educate excavators about natural gas lines to prevent damage.

Crowded utility corridors

Digging under city streets or sidewalks reveals an ever-increasing maze of pipes, wires, cables, and the like. As more underground utilities are packed into the limited space, cost and risk both increase. Mitigating risk calls for increased hand excavation, driving up costs. Managing crowded corridors will continue to be a key challenge for natural gas distribution companies and others who occupy this underground space.

Aging infrastructure

While the risk posed by natural gas lines and facilities is very low, some of the accidents are horrible and spectacular. The majority of local distribution lines and facilities have been installed since in the 1970s, but some were installed in the 1920s and 1930s, or even earlier, leading to a concern with aging infrastructure.

Leak-prone lines. A study released by the American Gas Association in 2012 found that approximately 9% of the distribution mains services in the United States are constructed of materials that are considered leak-prone. At the current pace of replacement, it will take up to three decades or longer for many operators to replace this infrastructure.[5] Leak-prone infrastructure is comprised of materials that are susceptible to corrosion or other material failure. It is also affected by soil environment, pressure, how the pipe has been maintained, its leak history, and even depth of burial and frost depth, owing to potential frost heave problems. The remaining 91% of the miles are almost 475,000 miles of protected steel and a little less than 645,000 miles of plastic pipe, according to the study.[6]

On the surface, the answer to the challenge of leak-prone infrastructure seems straightforward, at least from a public safety perspective: replace all leak-prone pipelines, and do it now. However, much of this infrastructure was installed years ago when population densities were lower. Over time new and larger buildings were constructed, leading to crowded, busier streets and sidewalks. As a result, replacement would be disruptive and expensive, not to mention potentially destructive to other underground utilities as city streets and sidewalks are torn up. The need for replacing lines must be balanced with ensuring affordable energy bills for customers, which is not an easy task.

Local distribution companies have developed and are implementing systematic programs to replace or rehabilitate all of their leak-prone pipe on a risk-ranked basis. Thus they replace the highest risk pipe and work toward the lower risk pipe as the program progresses. Managing distribution integrity is a major focus for natural gas distribution companies.

The challenge of replacing or rehabilitating leak-prone infrastructure gas pipelines has resulted in innovative techniques. These include robotic repairs, installation of rehabilitation liners, bursting the pipe and pulling plastic pipe into the old pipe as a replacement, and various new or improved technologies. Over time, legacy leak-prone infrastructure will be replaced or rehabilitated.

Of course, any releases presenting an immediate hazard are repaired or replaced as soon as they are discovered. Releases not presenting an immediate hazard are repaired or replaced on a prioritized basis.

Plastic pipe. In the 1970s, the industry switched to primarily high density and medium density polyethylene (HDPE and MDPE) pipe, presenting another challenge, as no one really knows the degradation rate of plastic. Some of the early plastic pipe was not HDPE or MDPE, and there were some failures. A number of organizations concerned with public safety have come together to address these failures and understand the performance of plastic pipe. These organizations include the American Gas Association, American Public Gas Association, Plastics Pipe Institute, National Association of Regulatory Utility Commissioners, National Association of Pipeline Safety Representatives, National Transportation Safety Board, and the Department of Transportation. These organizations participate on a committee that collects and assesses in-service plastic piping material failures with the objective of identifying possible trends in the performance of these materials.

Collected data include both actual failures and negative reports (forms that indicate that no failure occurred during the month). According to the AGA, it "has collected this data on behalf of the committee since January 2001, and the data is examined each time the committee meets in an effort to identify trends in the performance of plastic piping materials."[7] *Plastic Piping Data Collection Initiative Status Report* is published by the Plastic Pipe Database Committee of the AGA, and the latest report is available on the AGA Web site.[8]

Other public safety challenges

Another public safety challenge concerns vehicle drivers. Drivers may be under the influence of alcohol or drugs, be distracted, have a medical issue, or are speeding when they run into aboveground facilities, such as regulator stations or meter sets.

In an attempt to do the right thing and keep consumer rates as low as possible, consumer advocates sometimes challenge pipeline safety initiatives that are not mandated in regulations. They object to including funding for proactive safety practices in the rates. Current risk-based integrity management programs, however, provide LDCs a framework for explaining these proactive measures.

Reliability

Natural gas distribution systems have little room for error. When service stations run out of gasoline, motorists simply go to the station across the street. In contrast, houses, schools, commercial establishments, hospitals, and most other natural gas consumers have only one gas connection and no backup fuel. As gas pressure at the burner tip falls, the pressure can reach a point where the gas concentration no longer supports combustion, and the flame goes out, inconveniencing the customer and causing potential safety risks. Gas must reach each and every consumption point, even the ones at the far reaches of the distribution system, in the prescribed pressure range.

Likewise, appliances are designed to operate in a given energy range (measured in British thermal units, calories, or kilowatts) for maximum efficiency. Gas supplied with energy contents outside that range can be inefficient or even dangerous. Supply and interchangeability, then, are two key challenges.

Supply reliability at the city gate

To ensure customers get the gas when they need it, the gas must first be available reliably at the point of entry into the LDC system; most commonly, this is the city gate. Availability means the gas must be purchased and then transported. To that end, LDC supply departments purchase natural gas from a variety of sources and contract for transportation with several natural gas transmission operators. They thereby lessen exposure to any disruption from any one supplier and also spread out the transportation risk.

When distribution companies must depend on monopolistic suppliers and transmission lines controlled by governments, grave geopolitical supply and pricing risks arise. European LDCs know this from past experience.

Transportation capacity. Sufficient natural gas transportation capacity is critical to supply reliability, so LDCs support construction of transmission lines, interconnects, storage, and imports. They often participate in the politics, posturing, land rights issues, and theatrics involved with permitting those projects. Natural gas pipeline construction presents a challenging environmental dilemma. On one hand, natural gas is one of the most environmentally friendly fossil fuels. On the other hand, pipelines disrupt the land during construction.

Production availability. Pipelines carry natural gas from the point of production or import to the market. So in addition to sufficient supply transportation capacity, sufficient production capacity is critical. Production capacity is not a problem in countries like the United States, with its current supply of "shale gas." However, supply is currently a challenge in some countries.

One only needs to recall the late 1990s to realize supply can change rather unexpectedly. At that time the United States was planning LNG import facilities, which have now been recast as export facilities, driven by US unconventional natural gas production.

Storage. Natural gas production is relatively smooth and ratable over the year, but consumption is not. It varies between seasons and during the 24-hour period. Thus natural gas is placed into storage in times of lower demand and is withdrawn in times of higher demand. Sufficient working storage with the required injection and withdrawal rates ensure the ability to balance supply with demand.

Interchangeability

According to the American Gas Association, *interchangeability* is "a measure of the degree to which combustion characteristics of one gas are compatible with those of another gas. Two gases are said to be interchangeable when one gas may be substituted for the other gas without interfering with the operation of gas burning appliances or equipment."[9]

The UN Economic Commission for Europe defines it as "the ability to substitute one gaseous fuel for another in a combustion application without materially changing operational safety, efficiency, performance or materially increasing air pollutant emissions."[10]

Energy content. The energy content of a natural gas stream varies based on production reservoir characteristics, processing plant design and operation, and additional processing or blending (which may or may not be performed prior to the point of consumption). Ethane price and value changes can result in either extracting nearly 100% of the ethane from the methane stream or leaving it all in, drastically impacting stream energy content. Since appliances, power generation turbines, and other natural gas–fired equipment are designed to operate within a defined energy range, the natural gas industry closely monitors natural gas interchangeability.

Various organizations, such as the European Association for the Streamlining of Energy Exchange (EASEE), among others, attempt to develop and promote the simplification and streamlining of both the physical transfer and the trading of gas across Europe. Gas interchangeability is one of their focus areas.

In the mid- to late-1990s and early 2000s, NGL import facilities were being designed, and gas energy content interchangeability was a concern. This concern has lessened considerably with the advent of unconventional natural gas production in the United States.

Impurity specifications. Allowable limits for hydrogen sulfide, total sulfur, mercaptan sulfur, oxygen, nitrogen, carbon dioxide, mercury, water vapor, and perhaps other impurities or contaminants can also vary between regions, countries, and even within countries, presenting a barrier to global trade. Associations, and in some cases governments, work together to standardize natural gas specifications.

The UN Economic Commission for Europe in a recent report concludes, "The global harmonization of traded LNG quality is unlikely, even considering the 2030 timeline, because of the different economic and political forces on different consuming nations. However, some harmonization is both feasible and likely, along the lines of the regional characteristics of USA, Europe and Asia Pacific and some degree of harmonization should be enough to improve the growing world development of the LNG market."[11]

LDC system reliability

The first part of the reliability challenge is to ensure that gas is delivered at the city gate in sufficient quantities, at correct pressures, in the required energy content range, and with limited impurities. From there, the gas mains and service lines must move the gas to the ultimate users without interruption.

Pressure profile. As gas flows from the city gate toward the most distant customer, the pressure in the gas stream decreases due to friction of the molecules rubbing against the pipe wall and against each other. On colder days, more gas flows, which results in more pressure loss between the city gate and the customer. Since gas consumption is higher on cold days than on warm days, it stands to reason that on the coldest days, pressure loss is greatest (because flow rate is greatest). When natural gas distribution systems are first constructed, they are designed to move gas on the coldest day to the current and planned customers, perhaps with some spare capacity for the future. Pipe diameters and wall thicknesses are chosen to ensure sufficient pressure remains in the system at the most distant customer. As more consumption points are added between the city gate and the most distant customer, flow rate can increase to the point where the pressure at the most distant point drops to zero (fig. 14–2).

The same may happen if lines are extended to more distant customers and the gas flows farther, losing pressure as it goes.

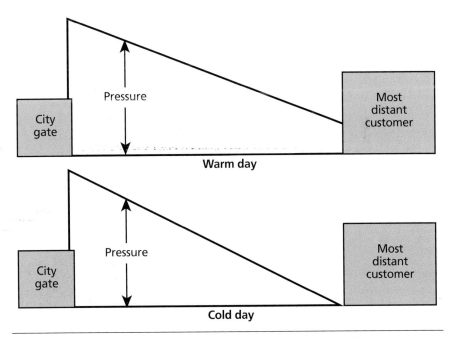

Fig. 14–2. Pressure profile. The slope of the line shows the pressure loss over a distance. The distance between the sloping line and the bottom line is the amount of pressure remaining in the line as gas moves along. On the warm day, the slope is not as steep as on the cold day, because the flow, and therefore pressure loss, is less. On the cold day, the friction loss means the pressure drops to zero, meaning gas never reaches the most distant customer.

Reinforcing the system. To ensure system reliability for even the most distant customer on the coldest day, engineers regularly review demand and pressure levels. They reinforce the system by replacing smaller diameter pipes with larger diameter pipes to decrease flow velocities and thereby friction loss. They replace lower pressure lines with higher pressure lines, construct parallel mains to split flows, thereby reducing velocity, and reroute and reconnect lines to shorten flow distances. Often they use a combination of several approaches; sometimes additional city gates are added.

Reinforcing the system to ensure adequate pressure and supply leads to two primary challenges: funding and construction. Typically demand growth occurs in small increments over time, whereas reinforcement expenditures happen in larger increments, meaning small demand (and therefore revenue) increases may require larger expenditures than can be economically justified. Installing reinforcement projects generally means excavating in congested areas and installing additional underground utilities into already crowded space.

Peak shaving. As a last resort, various *peak-shaving* techniques are employed to shave the peak off of the demand. Peak shaving includes some storage, for example liquefying natural gas and storing it for later gasification and injection into the system. It could include injecting a propane and air mixture into the system blended to an energy content consistent with system energy content, or storing compressed natural gas and releasing it into the system at peak demand times. Peak-shaving facilities are typically economically justified on the basis of avoiding investment in other facilities rather than on the basis of the revenue earned on the small incremental transportation of natural gas on the coldest day.

Efficiency

Investment and cost efficiency can be a bit of a dilemma for natural gas distribution companies. On the one hand, their rates are regulated, essentially capping the revenue they can earn. On the other hand, rates generally include return on investment and return of expenses. Ideally, then, investments and expense forecasts included in the rate calculations would be as high as possible, pushing up rates, but then the investments and expenses actually incurred would be as low as possible, driving up earnings. LDCs must balance these factors to balance the customer's desires with the owner's expectations.

Energy efficiency presents another dilemma. Local distribution companies derive revenue from moving gas, so they have a natural bias to encourage consumption. However, their direct connection with the customer puts them on the front line of encouraging conservation of energy. Natural gas distribution companies work with local regulators to help them understand and respond to the conservation challenge.

The industry and individual companies work to find innovative rate designs. For example, the American Gas Association suggests the following on its Web site:

> *"Decoupling" of utility revenues from natural gas sales is one method of removing such energy efficiency market impediments. Other regulatory incentives that would empower utilities to provide energy efficiency programs including non-volumetric and other innovative rate designs; recovery of energy efficiency program-related costs; recovery of revenue and margin losses associated with implementing energy efficiency programs; and performance-based incentives for utility shareholders and/or ratepayers that attain specific success metrics (such as predefined energy savings and/or cost efficiency targets).*[12]

Environmental Performance

Natural gas is widely viewed as the cleanest burning fossil fuel. When natural gas burns, it emits mostly water vapor and CO_2, and very little, if any, sulfur, nitrogen, or particulates, making it the most favored of fossil fuels. Methane is, however, a greenhouse gas (GHG), so emissions are closely monitored and tracked. Methane that would otherwise be emitted to the atmosphere can be captured by LDCs and other companies from organic waste sources, such as landfills or feedlots. It can then be cleaned and injected into the natural gas system and burned just like methane produced from gas and oil wells.

Aging Workforce

Finally, in the United States and many other developed countries, workers are retiring, taking with them years of accumulated knowledge and experience that is proving difficult to capture or replicate. The challenge of retaining knowledge developed through decades of experience spans the gamut of the other challenges.

Summary

- The overall challenge is to provide safe, reliable, efficient natural gas delivery with due consideration for the environment, while keeping stakeholders' needs balanced.
- Approximately 9% of distribution mains in the United States are constructed of materials that are considered leak-prone.
- The challenge of replacing or rehabilitating leak-prone infrastructure gas gave rise to innovative techniques.
- The AGA's Plastic Pipe Database Committee collects plastic pipe performance data.
- One of the leading causes of natural gas pipeline releases in North America and Europe is excavation damage.
- Gas must reach each and every consumption point, even the ones at the far reaches of the distribution system, in the prescribed pressure range.
- Sufficient natural gas transportation capacity is critical to supply reliability.
- Some harmonization of natural gas specifications is feasible and likely, along the lines of the regional characteristics.
- Reinforcing the system to ensure adequate pressure and supply leads to two primary challenges: funding and construction.
- Investment and cost efficiency can be a dilemma for natural gas distribution companies.
- Natural gas is widely viewed as the cleanest burning fossil fuel.
- Natural gas distribution workers are retiring, taking with them years of difficult-to-replace accumulated knowledge and experience.

Notes

1. Eurogas, *Statistical Report 2014* (Brussels: Eurogas, December 2014), 8. http://www.eurogas.org/uploads/media/Eurogas_Statistical_Report_2014.pdf.
2. US Energy Information Administration, "Natural Gas: Number of Natural Gas Consumers," *Independent Statistics and Analysis* (US EIA, December 31, 2014), http://www.eia.gov/dnav/ng/ng_cons_num_dcu_nus_a.htm.
3. Common Ground Alliance, "How 811 Works," http://call811.com/before-you-dig/how-811-works.

4. Common Ground Alliance, *Best Practices 12.0: Putting Great Ideas into Motion* (Washington, DC: CGA, March 2015), http://www.commongroundalliance.com/best-practices-guide.

5. American Gas Association, *Gas Distribution Infrastructure: Pipeline Replacement and Upgrades* (AGA Foundation, July 2012), ES-1, https://opsweb.phmsa.dot.gov/pipelineforum/docs/07-2012%20Gas%20Distribution%20Infrastructure%20-%20Pipeline%20Replacement%20and%20Upgrades.pdf.

6. Ibid., appendix A.

7. American Gas Association, "Plastic Pipe: AGA Viewpoint," https://www.aga.org/plastic-pipe.

8. Plastic Pipe Database Committee, *The Plastic Piping Data Collection Initiative Status Report*, American Gas Association (Washington, DC: PPDC, July 23, 2015), https://www.aga.org/sites/default/files/ppdc_july_2015_status_report.pdf.

9. American Gas Association, "Natural Gas 101, Interchangeability," *Knowledge Center*, https://www.aga.org/knowledgecenter/natural-gas-101/natural-gas-glossary/i.

10. UN Economic Commission for Europe, *Study on Current Status and Perspectives for LNG in the UNECE Region* (Geneva: UNECE, 2013), chapter 4, p. 5.

11. Ibid. 31.

12. American Gas Association, "Energy Efficiency: Regulatory Incentives for Utilities to Promote Energy Efficiency," *Policy Issues, Environment*, https://www.aga.org/energy-efficiency.

index

A

aboveground
    district pressure regulator with back-up monitor valve, 165
    pipelines, 33
    pressurized gas holders, 168–169
    storage facility, 302
    valves, 240
abrasion resistant overcoating (ARO), 90
absolute pressure, 60
absolute temperature, 60
absolute viscosity, 62
absorption, for dehydration of gas streams, 130–131
acceptable risk, 344
acetylene welding, 38–39
actuators
    for natural gas, 105–108
    oil as hydraulic fluid in, 107–108
    types of, 105–108, 122
adsorption, 130–131
advertising, 183
air compressors, 322
alarms. *See* indicators, alarms, and alarm filtering
Alberta Energy Regulator, 313
Allegro energy consulting, 229–230
allowed revenue, 338–339
alternating current voltage gradients (AC), 178
ambient heat, 200
ambient temperature, 290
American Gas Association
    decoupling recommendations by, 359
    definitions by, 63–64, 163, 355
    natural gas conversion by, 61
    plastic pipe statistics by, 80
    safety and land use regulations by, 45–46
American Petroleum Institute, 46, 86
American Society for Testing and Materials (ASTM International)
    list of six plastic pipe properties, 92, 121
    specifications for fusion by, 94
    standards for polyethelene plastic pipe by, 81
    tables by, for conversion of natural gas, 61
American Society of Mechanical Engineers (ASME International), 279
anodes, 118–119
ANSI gasket, for steel fittings, 96
API 5L, *Specification for Line Pipe* (American Petroleum Institute), 86
application interface, 267
aquifers, 119–120, 127, 168–169, 302
Aquitaine, Elf, 42
area pages, 262
asbestos, 89
as-builts, 331–332
asset condition determination, 244–245
asset integrity
    asset conditions for, 244–245
    data for, 227–231
    defects within, 223, 245–247
    distribution integrity management for, 224, 226, 353
    failure mechanisms and forces within, 231–236
    of local distribution pipeline operations, 176

management plans for, 242–244
preventing releases as, 236–242
risk within, 224–227
associated gas, 27
Association for Advancement of Cost Engineering (AACE International), 285
attack, 226
attribute data, 226, 331
auditing, 189
Auer, Carl, 23
augers, 321–322
automatic custody transfer (ACT) unit, 6
automatic shut-off valves, 295–296
average demand, 184

B

back-fed systems, 287
bad debt, 183
bagged off repairing, 34–35
ball check valves, 103–104
ball valves, 99–101, 103–106
barriers, protective, 240
Barrow and Company, 27
bars, 60, 219–220
Barton, Henry, 16
basic flow equation, 74–76
batches and batching, 7, 12
Beck, Samuel Adams, 25
bell and spigot joint, 32
Bernoulli's principle
applied to hydraulics, 67–69
applied to orifice meters, 112, 139
applied to water, 56–57
Best Practices (Common Ground Alliance), 351
bid packages, 186
the Big Inch Pipeline, 37
billing, 182, 336–337
biofuels, 2
biomass gas, 168
bleed valves, 100
block valves, 100, 295–296
Board of Gas Commissioners (Massachusetts), 25
boiling off, 200–201
book value, 346
booster stations, 12, 144. *See also* compressor stations
boring, 321–322
Boulton, Mathew, 16–17
Boyle's law, 63–64, 290

Bradford Glass Company, 28
branch connections, 319
breached lines, 245
British Gas Council, 194
British Standards Institute, 45–46
British thermal unit (Btu)
applied to gases, 133
definition of, 61
gases measured by, 37
bulldozer, 40
bundled rates, 336–337
Bunsen burner, 23
Bureau of Economic Geology, 202–203
butane
British thermal unit content of, 133
fractionation of, 131
gas processing plants removal of, 129
for liquefied natural gas, 136–137
removal for liquefied natural gas, 197
butt
fusion, 315–317, 320
welding, 93

C

calibration, 142
call centers, 180–181
One Call, 231, 236–239, 351
cap weld beads, 324
caps. *See* fittings
carbureted water gas, 20–21
carburizing gas furnaces, 17
Carlsbad, New Mexico explosion crater, 132
Carroll, Lewis, 15
cast iron pipe
corrosion of, 235, 242
for early distribution lines, 80
for gas pipelines, 22
for mains, 32–33, 163
removal of, 286–287
valves for, 28
cathodic protection, 118
caverns, for gas storage. *See* salt dome caverns
centipoises, 62
centistokes, 62
central control rooms, 174
booster stations for, 144
controllers of, 138–139, 146–147, 150–156
emergency responses by, 156–157
gas control console for, 126, 146

Index 365

local distribution companies, 252
local or central use of, 258
maximum allowable operating pressure in, 152
for natural gas transmission lines, 146-157
quality control in, 133, 153-154
scheduling for, 147-150
centrifugal compressors, 144
Charles' law, 63-65, 200
charters, 25
check valves, 99, 103-106
chemical feedstocks, 2
Chicago Utility Alert Network, 351
Churchill, Winston, 159
circumferential welding, 88
city border stations. *See* city gates
city gates
    facility design at, 303-304
    flow control at, 146
    human machine interface and, 262-263
    local distribution companies and, 145-146
    mains for, 315-318
    natural gas supply reliability at, 354-355
    stations at, 160-162, 255, 259, 303-305
City of London Gas Act of 1866, 25
clamps, 33, 98, 246, 323
Clegg, Samuel, 17, 22
Cleveland storage disaster, 41-42
close interval surveys (CIS), 178
coal. *See also* methane
    carbonization of, 19-21
    for gas, 17-19, 23, 37
    gasification of, 16-18, 168
    1900-1950 use of, 30-31
coal tar enamel (CTE), 87-89
coatings, 117-118
    fusion bond epoxy, 87-90, 122
    for natural gas distribution, 86-91
codes. *See* engineering standards and codes
coke, 19, 21
collars, for piping, 34
collections departments and bad debt, 183
color coding, for pipes, 82-83
combustible gas indicator (CGI), 216-217, 220
combustion, 156
command processing time, 271

commercial development, 23-26
Committee of Enquiry, 38
Common Ground Alliance (CGA), 230, 237-240, 351
company specific data, 231
comparable sales technique, 345
Competitive Advantage: Creating and Sustaining Superior Performance (Porter), 4
composite fittings, 91, 97-98
compressibility
    definition of, 63
    line packing and, 76
    practical operating point for, 66-67
compression
    bypassing by local distribution companies, 149
    fittings for, 94, 98, 247, 315
compressor stations, 12
    diaphragm actuator use in, 106-107
    friction loss prevention with, 71-72
    history of, 40
    operations for natural gas, 55
    types of, 142-144, 297-300
    US locations of, 125-127
compressors
    configuration for, 153
    for gas, 28-29, 40, 55
    for gathering, 134-135
    history of development of, 28
    Old Battleship, 29
    operation of, 55, 153
    pressure induction of natural gas by, 52
    refrigeration technology for, 41-42
    types of, 11-12, 143-144, 298, 322
computers
    for distribution, 43
    for flow, 109, 113-114, 140
    modeling programs for, 153
concrete coating, 90-91
condition data, 226
configuration tools, 266-267
confined space leak detection, 219
Confucius, 223
consequences of risk, 225
Consolidated Gas Company of New York, 25
consolidation, of natural gas, 25
consortia, for excavation incidents, 351-352
construction
    as-builts and inventories for, 331-332
    basics of, 311-314

of bridges for pipelines, 33
of facilities, 325–327
of gas transmission lines, 27–30, 38–40
inspections, safety, and quality control for, 327–331
for local distribution pipeline operations, 176–177
plastic lines and connections for, 314–320
steel pipe as lines for, 320–324
contaminate gases, 10–11
contaminate molecules. See contaminate gases
contract management, 189
control room operations, 138–139, 146–157, 265
control systems, 251–252, 266–267, 309
control systems and SCADA, 251–252
design and control of, 257–274
pressure regulator station schematic, 254
SCADA, 255–257
SCADA screenshot, 253
control valves, 174, 254, 266
controllers
in central control rooms, 138–139
monitoring by, 150–156
for natural gas lines, 125–127
operation plans for, 146–147
cooling and lubricating systems, 144
cooperatively owned local distribution companies, 335
de Coriolis, Gaspard-Gustave, 114
Coriolis force, 114–115
coriolis meter, 114–115, 139
corrosion
in 40 percent of preventing releases, 241–242
prevention of, 132–133, 177–178, 242
soap solution bubbling over, 212
types and factors of, 132, 178, 235
corrupted data, 269
cost of gas adjustment, 182
cost of service (COS) theory, 337–339
couplings. See fittings
critical pressure, 66, 196
critical temperature, 66, 196
Cronkite, Walter, 206
crude oil, 5–8, 26–27, 37
Cunningham, James, 40
current flow rate, 74
custody transfer unit, 5–6

customary system, US (USCS), 59
customer service, 180–183
cybersecurity, 273

D

data. See also SCADA
for asset integrity, 226–231
construction collection of, 331
corrupted, 269
historic database of, 267–268
real-time operating, 264–266
sharing for design and control, for control systems, 272
debt, 183, 338
decoupling, 359
defects and failures, of asset integrity, 223, 245–247
defense, 226
Degrees Rankin, 60–61
dehydration, 130–131
dekatherms, 61, 337
delivery operations, for natural gas, 144–146
delivery stations, 305–306
demand forecasting, 184
density, 61, 71–72, 291
Department of Transportation, US, 206, 229, 237
depleted fields, 119–120, 127
depleted reservoirs, 119–120, 127
design and control, for control systems, 257–274
design and engineering, 277
control system design for, 266–267, 309
natural gas engineering functions for, 278
pipeline design for, 286–296
of power systems, 308–309
process for, 280–283
standards and codes for, 279–280
design day, 184
destination points, 265
detailed design and engineering, 285–286
deviation factor. See Z factor
diameter ranges, 163, 166–167
diaphragm actuators, 106–108
diaphragm meters, 109–111, 115
diaphragm valves, 103, 254
diaphragms, 103, 106–111, 115, 254
differential pressure, 139–140, 287–288
Digger, 351

Index 367

direct current voltage gradients (DC), 178
direct volume meters, 109–111
disbonding of tapes, 90
discretionary expenditure decisions, 340–341
dispatch for customer service, 181
dispersion, 220
distance, 291
distribution, 2, 308, 336. *See also* local distribution companies
    from 1950 to 2000, 42–43
    coatings for, 86–91
    integrity management of, 224, 226, 353
    lines for, 13–14, 40, 80, 129
    mains for, 162–163, 329
    of manufactured gas, 30
    of natural gas, 23–25, 32–35, 53–57, 79, 80–122
    regulation of, 44–45
distribution integrity management plan (DIMP), 224, 226
distribution system operator (DSO), 336
district regulators, 164–165. *See also* regulator stations
ditches, 28–29, 40, 316, 318
double submerged arc welded seam method (DSAW), 84–86
downstream, processing and marketing, 4
Drake, Edwin, 26
Dresser coupling, 28
drilling
    horizontal directional, 43, 231–232, 320–321
    for natural gas, 18, 37
drips, 34, 133
dry gas, 129–130
dual fueled vessels, 201
dynamic pressure, 68
dynamic viscosity, 62

E

easements, 282, 314
Economic Commission for Europe, UN, 355–356
economic value, 133, 344–345
Edison, Thomas, 23, 51
efficiency, 358–359
811, 351
Eisenhower, Dwight D., 349
elastic limit, 63
elbows. *See* fittings

electric actuators, 108
electric resistance welded seam (ERW), 84
electrical lines, 232–233
electrical potential, 178
electrofusion plastic fittings, 93–94, 317
electrolysis, 117–118
elevation, 53, 72–74
emergency response
    as-builts for, 331–332
    by central control rooms, 156–157
    for local distribution pipeline operations, 178–179
energizing, of anodes, 119
energy, 1–2, 30, 52–53, 356, 358
    kinetic, 56, 66
    of natural gas, 109
    potential, 66, 76
Energy Information Administration, US, 120, 349
engineering. *See* design and engineering
engineering standards and codes, 279–280, 295
Engineering Standards Committee. *See* British Standards Institute
English system. *See* customary system, US
environmental performance, 359
epoxy solution, 42. *See also* fusion bond epoxy
equations of state, 66
equipment
    calibration of, 142
    as category of failure mechanisms and forces, 234
    definition of, 79
    for 40 percent of preventing releases, 241
    personal protective, 331
Ervia, 336
ethane, 129
    British thermal unit content of, 133
    fractionation of, 131
    for liquefied natural gas, 136–137
    price and value changes of, 356
    removal of, for liquefied natural gas, 197
ethylene, 20
ethyl-mercaptan, 128
European Association for the Streamlining of Energy Exchange (EASEE), 356
excavation incidents
    consortia for, 351–352
    as failure mechanisms, 231–232

One Call centers for, 237–238, 351
    public safety and, 351–352
    with releases, 236–240
excess flow valves (EFV), 167, 241
expected value, 341–342
expenditure decisions, 340–344
expenses, 339
Explorer Pipeline, 9
export terminals, for liquefied natural gas, 195, 202, 355
exposure, 225, 236
external corrosion, 235

F

fabricated pipe, 326–327
facilities, 145, 308
    aboveground storage, 168–169, 302
    construction of, 325–327
    functions for, 296–297
    hubs as, 12–13, 301–302
    interchange, 300–302
    for liquefied natural gas, 203, 355
    marine export, 7–8
facility design, 296–307
factories, and natural gas, 23–24
failed main, 232
failure mechanisms and forces, 231–236
farm taps, 168
feasibility, 280–281, 283
Federal Energy Regulatory Commission (FERC), 44–45, 148, 180, 203, 313
Federal Power Act, 36
Federal Power Commission, 45
Federal Trade Commission (FTC), 36
feeder mains, 162–163
FERC Order No. 436, 45
FERC Order No. 636, 45
field meter, 116
field operations, for natural gas, 138–144
fill rates, 120
filtering strategy, 270
finance and administration groups
    for local distribution pipeline operations, 189–190
first cut liquid (oil) pipelines, 3
fittings
    for meter sets, 326–327
    for repairing defects and failures, 246–247
    types of, 28, 91–98, 247, 315, 317–318, 320
fixed charges, 182
fixed interval reporting, 268–269
flame ionization detector (FID), 213–214
flame retardant clothing (FRC), 331
flanges. *See* fittings
flappers, 104. *See also* valves
flares, 207–208
flaring, 207–208
flashing, 194
floating storage regasification units (FSRUs), 202
flow, 98, 146, 188, 357
    computers for, 109, 113–114, 140
    excess valves for, 167, 241
    factors of, 288–289
    hydraulics and, 51–59, 67–76
    rates of, 111–112, 116, 154
    slug, 131–132
    summary pages for monitoring pipelines, 150–151
    velocity of, 291–293
flow charts, 130, 137, 147–148, 243
40 percent of preventing releases, 241–242
fracking, 3
fractionation
    of gases, 131
    for natural gas lines, 131–133
frequency. *See* update frequency
friction loss, 54–59, 71–72, 289, 291
front end engineering and design (FEED), 278, 283–285
front end loading (FEL). *See* front end engineering and design
front-end mounted optical methane detector (OMD), 214
function chart, for local distribution companies, 173
fungus-like growth, to detect leaks, 211
fusion, 315–320
fusion bond epoxy (FBE), 87–90, 122

G

gas. *See also* gas pipelines; governors; history, of gas
    associated, 27
    behavior of, 63–67
    carbureted water, 20–21
    chromatographs for quality control, 133, 153
    clouds, from leaks, 209
    from coal, 17–19, 23, 27
    compressors for, 11–12, 28–29, 40, 55, 143–144, 298, 322

control consoles for, 126, 146
governors, 22, 304
interchange points for, 144–145
interchangeability, with liquefied natural gas, 197–198
leak migration of, 210, 218–219
for lighting, 16–18, 26
measurement of, 173–175
processing plants for, 129–131, 135–137
production and consumption of, 3–14
regulation from 1950 to 2000, 43–46
regulation from 1900 to 1950, 35–38
regulation in industrial Revolution, 24–26
sampler for natural gas field operations, 140–141
"sniffers," 156
storage of, 40–43, 119–121, 127, 168–169, 198, 302
supply management of, 171, 184–190, 354–355
transmission lines, 10–14, 27–30, 38–40, 135–136
velocity of, 288
Wobbe index for, 198
Gas Act of 1948, 38
Gas Light Company of Baltimore, 18
gas pipelines, 3, 40, 79
development of, 24–25
interconnection points for, 144–145
materials for, 21–22
Gas Regulation Act of 1920, 38
Gas Safety Management Regulations (GSMR), 230–231
gasification, of coal, 16–18, 168
gaskets, 32, 95–96
Gas-Works Clause Act, 25
gate valves, 99, 102–103, 105–106
Gates, Bill, 251
gathering, 2, 7
natural gas lines for, 134–135
pipelines for, 5–6, 14
gauge pressure (psig), 60
Gay-Lussac, Joseph Louis, 63, 290
Gay-Lussac's Law, 63–65
Gazprom, 149, 336
geographic information systems (GIS), 175–176, 251–252
geospatial data, 226, 331
geospatial technology, 175
Gesner, Abraham, 23

girth welding, 87–89, 91, 330
globe valves, 99, 101–102, 106
governors, 22, 304. *See also* regulator stations
Great Western Iron Company, 23–24
greenhouse gas (GHG), 359
grids, 258, 287
ground bed, 118
group navigation pages, 260–261
gun barrels, for pipelines, 21, 79

H

Hart, William Aaron, 18
Hayworth, Geoffrey, 38
Health and Safety Executive, 230–231, 233–234, 280
heat, 16, 31, 62–63, 200
heating plate, 315
heavy molecules, in natural gas, 131
Hebden Bridge mill, 17
Henry Hub, 137, 301
high density polyethylene (plastic) pipe (HDPE), 82–83, 353
high pressure transmission pipelines, 71
highest and best use technique, 345
high-pressure mains, 162–163, 166, 304–305
historic database, 267–268, 272
history, of gas
during Industrial Revolution, 23–30
for liquefied natural gas, 194
for manufactured gas, 16–21
for natural gas, 18, 27–30
from 1950 to 2000, 42–46
from 1900 to 1950, 30–42
horizontal directional drilling (HDD), 43, 231–232, 320–321
horse drawn drip wagon, 34
Horsely Iron Works, 22
hot tapping, 246–247
house meter assemblies, 13, 95, 167, 254
hubs, 12–13, 137, 301–302. *See also* interchange facilities
human decisions, machine *vs.*, 257–258
human machine interface (HMI), 126, 260–264, 271
hydrant system, 54
hydraulic gradients, 288, 291
hydraulic modeling software, 296
hydraulics
friction loss and, 54–59, 71–72
for pipelines, 59–77

hydrocarbon fluids, properties of, 59 67
hydrocarbons, 129–130, 213–217
hydrostatic testing, 329

I

ideal gas, 63–66
Imperial Continental Gas Association, 22–23
import terminals, 136, 202, 355
impurity specifications, for interchangeability, 356
incident reporting, 229
indicators, alarms, and alarm filtering, 270–271
Industrial Revolution, 23–30
inference meters. *See* inferential meters
inferential meters, 109, 111–115, 139–140, 145, 265
information density, 260
infrastructure, aging, 352–353
injection molded plastic fittings, 92–93
injection point, 7
insects, to detect leaks, 211
inspections, 155, 177
    for construction, 327–331
    internal line, 244–245
installations
    keyhole, 322
    of mains, 164, 177
    processes for SCADA, 262
    trenchless, 43
integrity management, 236, 243–244
    as-builts for, 331–332
    of distribution, 224, 226, 353
integrity management plans (IMPs), 236
integrity surcharges, 340
interchange facilities, 300–302
interchangeability
    of gas, 197–198
    natural gas systems and, 355–356
interconnection points, 144–145, 300–302
interconnects. *See* hubs
intermediate compressor stations, 297, 299–300
intermediate pressure mains, 162–163, 304–305
internal corrosion, 132–133, 235
internal line inspection (ILI), 244–245
internal rate of return (IRR), 342
International Cost Engineering Council, 285
International System of Units (SI), 58

Interstate Commerce Commission, 45
interstation mains, 162–163
inventories, for construction, 331–332
invoicing systems, 115
iron content, of steel pipe, 235
isolation valves, 295–296

J

Jefferson, Thomas, 15
Johnson, Lyndon, 179
joint clamps, 33
Joint Utility Locating Information for Excavators (JULIE), 351
joints. *See* steel pipe
joule, 67
Joule, James Prescott, 67
Jules-Thompson valves, 194

K

Kelvin scale, 60–61
kerosene, 23, 26
keyholes, 322
kinematics, 62
kinetic energy, 56, 66

L

Lacq gas field, 42
laminar flow, 69–71
landfill gas, 168
laser devices, for leak detection, 216
law of conservation of mass and energy. *See* thermodynamics
leak clamp, 98
leaks, 246
    detection instruments for, 213–220
    detection methods for, 211–212, 330
    epoxy solution for prevention of, 42
    management of, 207–220, 330
    in pipelines, 34–35, 37–38, 42–43
    prone lines for, 352–353
    service line discovery of, 245
    surveying of, 175, 330
lean oil, 130
LeBon, Philippe, 17
length of pipe, for pipelines, 54–59
life-cycle cost, 281
lifters, 40–41
light
    history of gas for, 16–18, 26

manufactured gas for, 30–31
Linde, Karl Von, 194
line operation optimization, 152–153
line pack, 121, 170
    definition of, 75–76
    packing and unpacking, 76, 148–150, 152
line pipe, 83–84, 86. *See also* steel pipe
linear transmission lines, 287
liquefied natural gas (LNG), 41–43, 120, 168–171, 194
    creation of, 196–202
    export terminals for, 195, 202, 355
    facilities in US for, 203, 355
    import terminals for, 136, 202, 355
    peak shaving liquefaction operations for, 198–199
    pretreating of, 197
    processing of, 136–137
    storage and transportation for, 199–201
the Little Big Inch Pipeline, 37
live repairing, of pipelines, 34
local distribution companies (LDCs), 115, 149, 153, 252
    business models and decisions for, 336–346
    buying and selling of gas by, 128–129
    city gates and, 145–146
    master meter systems for, 335–336
    natural gas systems and, 13, 356–358
    ownership of, 44, 121, 160, 335–336
    pipeline operations for, 159–190
    safety and land use regulations for, 46
local distribution lines, 13–14, 129, 160–161, 177–178
local distribution pipeline operations, 159, 183, 320
    construction for, 176–177
    emergency response for, 178–179
    finance and administration groups for, 189–190
    gas measurement for, 173–175
    maintaining service, 170–172
    supply management for, 184–189
    systems within, 160–172
local distribution system elements, 121
local station SCADA, 256
localized control rooms, 174
locating and marketing of excavation incidents, 238
Lodge, Henry, 17
log pipelines, 28
logs for control rooms, 272
longitudinal weld seam, 84
Lowe, T. S. C., 20
lower explosive limit (LEL), 210–211, 217
lower flammable limit (LFL), 210
low-pressure mains, 162–163, 304–305
low-voltage distribution system, 308
Luther, Martin, 311

M

M. Ambroise & Company, 17
magnesium anode, 119
Magnolia Gas of Dallas, 40
mains
    cast iron pipe for, 32–33, 163
    for city gates, 315–318
    for local distribution pipeline operations, 162–164
    plastic, 164, 177, 315–318, 320
    types of, 162–163, 166, 304–305, 329
maintenance precautions, 155–156
management of change (MOC) process, 328
management plans, 236
    for asset integrity, 242–244
    for distribution integrity, 224, 226
Manby, Aaron, 22
mandrel, 84
mantle, 23
manual actuators, 105
manual operators. *See* actuators
manual regulators, 174
manual valves, 295–296
manufactured gas
    conversion away from, in US, 37–38
    discovery and early use of, 15–23
    natural gas compared to, 23–24
    1900 to 1950 use of, 30–42
    regulation of, 35–38
maps, 150–151, 238–239
marine export facilities, 7–8
market centers, 7. *See also* hubs
marketing, 4, 27, 183, 238
master meter systems, 335–336
master meters, 116, 122
maximum allowable operating pressure (MAOP), 152, 293–294, 304
McLeod, J. Earl, 40
measurement
    of gas, 173–175
    of hydrocarbon fluids, 59–61

of manufactured gas, 37
for natural gas, 37, 139–142
of pressure, 59–60, 109, 139–140
of temperature, 60–61, 109, 139–140
medium density polyethylene (plastic) pipe (MDPE), 353
medium-pressure mains, 304–305
Melville, David, 18
membrane tanks, 201
membranes. *See* diaphragms
meter factor, 109
meter index, 115
meter provers, 79, 116, 122
meter sets, 326–327
meter station, 325
meter station graphics page, 152
meters, 152
    at city gates, 146
    customer service reading of, 181
    delivery by, 9–10
    diaphragms for, 109–111, 115
    excess flow valves for, 167, 241
    factor for, 109
    for gathering, 135
    house, assemblies for, 13, 95, 167, 254
    provers for, 79, 116, 122
    risers for, 320
methane
    for coal carbonization, 20
    conversion into liquid, 131
    diagram of molecule of, 129
    after gas processing, 11
    as greenhouse gas, 359
    leak concentration of, 210
    leak detection of, 213–217
    for liquefied natural gas, 196
    natural gas content of, 133
    in raw gas, 140
Methane Pioneer, 42, 194
methyl-mercaptan, 128
metric system, 58
Metropolis Gas Act, 25
midstream, 4
Miesner, Ella, 277
migration, of gas leaks, 210, 218–219
mimic boards, 262
Ministry of Fuel and Power, 38
mitigation
    for integrity management, 236
    of risk, 226, 351
monitor, for district regulators, 164–165
monitor screen, 262, 264
monitor valves, 254

monitoring pipelines
    from central control rooms, 150–152
    by district regulators, 164–165
monitoring routes, 239
monopolies, of natural gas, 35
Montefiore, Moses, 22–23
Moss Maritime, 201
Moss tanks, 201
Muhlbauer, W. Kent, 225–226
municipal ownership of local distribution companies (LDC), 160
municipal water systems, 53–57
Murdoch, William, 16–17, 19

N

National Balancing Point (NBP), 137
National Energy Board (NEB), 313
National Gas Archive, 38
National Transportation Safety Board, US, 206
natural gas. *See also* methane
    actuators for, 105–108
    British thermal unit of, 37
    combustion of, 156
    commercial development of, 23–26
    compressor operations for, 52, 55
    definition of, 2–3
    distribution, components of, 79–122
    distribution compared to municipal water systems, 53–57
    distribution of, 23–25, 32–35, 53–57
    drilling for, 18, 37
    field operations for, 138–144
    flaring of, 207–208
    future challenges, 350–359
    for heat, 31
    history of, 18, 27–30
    house meter assemblies for, 13, 95, 167, 254
    interchange facility design, 300–302
    lines for, 125–146, 232–233
    liquids, 2, 10–11
    methane content of, 133
    monopolization of, 35
    in pipelines, 9–14
    portable compressed, 171
    positive displacement compressors for, 11–12, 143, 298
    pressure and, 52–53, 61, 63, 66–67
    processing plants for, 11, 22–23, 25
    regulation of, 24–26, 35–38, 43–46

Index 373

science of, 51, 59, 61–62, 109
storage of, 13, 76, 120, 127, 302–303, 355
supply portfolios for, 186–187
systems, 354–358
transmission line operations for, 125–157
transmission lines for, 9–14, 18, 27–30, 63, 80, 135–136
transmission lines welding for, 38–39
US consumption by month of, 74–75
US conversion towards, 37–38
the US pipeline network of, 126
value chain for, 14, 128
water vapor method for, 130–131
winter demand of, 186
Natural Gas Act, 36, 45
Natural Gas and Natural Gasoline Division, 37
natural gas distribution systems, components of
actuators, 105–108
anodes, 118–119
coatings, 86–91
fittings, 91–98
meter provers, 79, 116, 122
meters, 109–115
odorant skids, 117
pipe, 80–86
rectifiers, 117–118
storage, 119–121
valves, 98–105
natural gas lines
controllers for, 125–127
electrical lines and, 232–233
fractionation for, 131–133
gathering, 134–135
liquefied natural gas using, 136–137
for natural gas transmission line operations, 128–146
for pipeline hubs, 137, 301
Natural Gas Pipeline, 40
Natural Gas Pipeline Safety Act, 46
Natural Gas Pipeline Safety Act of 1968, 179
Natural Gas Policy Act, 45
natural gas transmission line operations, 125–127
central control rooms for, 146–157
natural gas lines for, 128–146
natural gas transmission lines, 18, 135–136
central control rooms for, 146–157
compressors for, 11–12

hubs for, 12–13
Industrial Revolution building of, 27–30
local distribution lines compared to, 129
local distribution lines for, 13–14
pipe for, 80
raw gas for, 10–11, 129–130
vapor pressure inside of, 63
negative buoyancy, 90–91
net present value (NPV), 344
New London, Texas school explosion, 46, 205–206
Newton, Isaac, 63
"nice to know" data, 268
nighttime leak repair, 179
1950 to 2000
history of gas from, 42–46
1900 to 1950
history of gas from, 30–42
nominations, for scheduling, 147
nonhazardous gas leaks, 207–208, 246
nonparametric data, for real-time data, 265
nonreturns, 343–344
North American Energy Standards Board (NAESB), 148, 188
Northeast Gas Association (NGA), 244

O

odor, for detecting leaks, 211
odorant injection rate, 262–263
odorant skids, 117
odorization facilities, 145, 175
oil, 2
as hydraulic fluid for actuators, 107–108
tanks for, 26
transmission pipelines for, 40
types of, 3, 5–8, 26–27, 37, 130
Oil Transportation Association, 138
Old Battleship compressor, 29
One Call, for safety
establishment of, 236
for excavation incidents, 237–238, 351
information about, 231
systems for, 239
one-way feeds, 170–171, 287
open access rates, 336
operating pressures
maximum allowable, 152, 293–294, 304

for pipelines, 81–82, 294–295
operations, of local distribution pipelines, 175–180
operations personnel, 176–177
operators. *See also* actuators
    for field operations, 139, 142
    for leak detection, 213, 218
optical methane detector (OMD), 214–215
organizations, for gas distribution, 35
orifice meters
    Bernoulli's principle and, 112, 139
    for gathering lines, 9–10
    inferential, 112–113, 139–140, 265
origin compressor stations, 297–298
origin points, 265
origination compressor stations, 142–144, 297–298
outside force damage, 240–241
ownership models, of local distribution companies, 44
    cooperative, 335
    for facility storage, 121
    master meter, 335–336
    municipal, 160

P

packing the line. *See* line pack
Pall Mall, 17
parameter values, for real-time data, 265
Pascal, Blaise, 60
pascals (Pa, kPa), 60
peak shaving, 170, 198–199, 358
permits, 313–314
personal protective equipment (PPE), 331
Petroleum Administration for War (PAW), 37
petroleum products terminals, 8–9
Philadelphia, 31, 33
The Philadelphia Company, 24
Phillips & Lee's mill, 17
piercing, 322
pigs, for valve cleaning, 101
pipe. *See also* pipelines; plastic pipe; steel pipe
    characteristics of, 81, 293
    color coding of, 82–83
    diameter of, pressure loss due to, 289
    industry use of, 43
    length for pipelines, 54–59
    for natural gas transmission lines, 80
    probability of failure triad, 225–226

    selection of, 293–295
    stressing and straining of, 81–82
    types of, 82–83, 326–327, 353
    welding for, 84–86
Pipeline and Hazardous Materials Administration, US, 206–207
pipeline design, 286–289, 291, 293–296
pipeline networks. *See* grids
pipeline value chain. *See* value chain
pipeliners, 27, 40
pipelines
    aboveground, 33
    basics of, 1–3
    design of, 286–296
    ditches for, 28–29, 40
    flow within, 51–59
    for gathering, 5–6, 14
    grids of, 258, 287
    hubs for, 137, 301
    hydraulics of, 59–77
    interconnection points for, 144–145, 300–302
    leaks in, 34–35, 37–38, 42–43, 352–353
    log, 28
    municipal water systems use of, 53–54
    for natural gas, 9–14
    for oil, 5–8, 27, 37
    operating pressures of, 81–82, 294–295
    regulation of, 26, 45
    repair work of, 34–35
    stakeholders of, 281–282
    State of the National Pipeline Infrastructure, 229–230
    Tennessee Gas Pipeline, 46
    trenchless construction of, 43
    value chain for, 5
Pipelines Safety Regulations of 1996, 280
pipe-to-soil potential, 178
piston actuator, 106–108
Pittsburgh, 23–24, 27
Pittsburgh Brewing Company, 27
Pittsburgh Plate and Glass (PPG), 27
planning and design for excavation incidents, 237
plastic. *See also* plastic fittings; plastic pipe
    lines and connections, 314–320
    mains, 164, 177, 315–318
    service lines inside steel lines, 242
plastic fittings, 92–94, 98, 317

Index 375

plastic pipe
    defective, 234–235
    design and engineering of, 285
    infrastructure, aging and, 352–353
    operating pressures of, 294–295
    polyethylene, 43, 82–83, 232–234, 353
    steel displacement by, 80–81
    steel plpe compared to, 295
Plastic Pipe Database Committee, 353
Plastic Pipe Institute, 83
Plastic Piping Data Collection Initiative Status Report by Plastic Pipe Database Committee, 353
plug valves, 99–100, 105–106
plugs, 246–247
pneumatic actuators, 106–108, 122
points, for data, 265–266
Poiseuille, Jean Louis, 62
polling, 268
polyethylene plastic pipe (PE), 43, 82–83, 232–234, 353
    mains for, 164
portable compressed natural gas (CNG), 171
portable flame ionization detector (FID), 213
Porter, Michael
    Competitive Advantage: Creating and Sustaining Superior Performance by, 4
positive displacement compressors, 11–12, 143, 298
positive displacement meters. *See* direct volume meters
potential energy, 66, 76
pounds per square inch (psi), 59–60
power companies, 35–36
power system design, 308–309
pressure. *See also* measurement; pressure loss
    aboveground district pressure regulator with backup monitor valve, 165
    American Gas Association definition of, 163
    Bernoulli's principle and, 56–57
    booster stations for, 12
    city gate stations and regulators for, 145–146
    control of, 146, 254, 266
    drop, 254, 288–289
    flow related to, 51–53, 55, 291–292, 357
    governors for, 22, 304
    high pressure transmission pipelines, 71
    local distribution companies and, 153, 173–175, 357
    mains for, 162–163, 166, 304–305
    material considerations and, 295
    maximum allowable operating, 152, 293–294, 304
    metering and reduction station for, 325
    monitor screen for human machine interface, 262, 264
    per square inch in pipes, 73–74
    pipeline design and, 287–288
    quality control testing of, 330
    rate changes, abnormal of, 154
    reduction valve for, 162
    in regulator stations, 254, 304–305
    regulators for, 145–146, 164–165, 174, 258, 313
    summary pages of, for monitoring pipelines, 150–151
    testing, 155, 329
    types of, 60, 63, 66–68, 139–140, 196, 287–288
    of water, 72–73
    weather and, 357
pressure container. *See* sampler
pressure control, for local distribution pipelines, 173–175
pressure loss, 289–293
pressure reduction station, 325
pressure regulators. *See* district regulators; governors
pretreating liquefied natural gas, 197
prevention and management steps for failure, 226
price controls, 45
primary provers, 116
probability of failure (PoF) triad, 225–226
probes, 219
process and instrumentation diagram (P&ID), 265
processing plants
    for gas, 129–131, 135–137
    for liquefied natural gas, 136–137
    for natural gas, 11, 22–23, 25, 129–130
production processing facility, 5
project analysis, 342
propane
    Btu content of, 133
    fractionation of, 131

gas processing plants removal of, 129
liquefaction of, 136–137
for liquefied natural gas, 170
removal for liquefying natural gas, 197
provers, 79, 116, 122
public data sources, 229–231
public education for excavation incidents, 239
public relations for local distribution pipelines, 183
public safety, 350–354
public service commissions, 36
public utility commissions, 36
Public Utility Holding Act (PUHA), 36
purchased gas, 182, 340
purchasing, for gas supply management, 184–187

Q

quality assurance (QA). *See* quality control
quality control (QC)
in central control rooms, 133, 153–154
for construction, 327–331

R

raised face weld neck flanges, as steel fittings, 95–96
Rankin scale. *See* Degrees Rankin
rate base, 338
rate of return, 338, 342
rates, for finances, 337–339
by finance and administration groups, 189–190
for transportation and storage for purchasing, 186
Ravitch, Diane, 193
raw gas, 10–11, 129–130, 140
reading meter, 115
real gas law, 66
real-time data, 264–266
reamers, 321
reconstruction cost new, 346
rectifiers, 117–118
reduced port ball valves, 101
reducers. *See* fittings
reduction valves, 162
refined products, 7–9, 37
refineries, 7–9

refrigeration technology
compressors using, 41–42
in gas processing plants, 131
liquefied natural gas from, 199
storage, 168–169
regasification, 202
regulation, 239
Gas Safety Management Regulations, 230–231
during Industrial Revolution, 24–26
from 1950 to 2000, 43–46
from 1900 to 1950, 35–38
Pipelines Safety Regulations of 1996, 280
regulation points, for pressure, 292
regulator stations. *See also* district regulators; governors
pressure in, 254, 304–305
regulators. *See* district regulators; governors; regulator stations
regulatory compliance, 179–180
relative density. *See* specific gravity
releases
management of, 207–208
prevention of, 236–242
reliability of local distribution companies (LDCs), 356–358
remote methane leak detector (RMLD), 215–216
remotely controlled valves (RCVs), 295
repair work for pipelines, 34–35
repairing defects and failures, 246–247
replacement cost, 346
report by exception, 268–269
reporting and evaluation, of excavation incidents, 240
reports and logs, for control systems, 272
reservoirs, for storage, 3, 168–169, 302
depleted, 119–120, 127
retort, 19
returns, 338, 341–342
revenues, 337–340, 357
Reynolds, Osborne, 71
Reynolds number, 71
rich oil, 130
right-of-way (ROW), 142, 190, 282, 313–314, 345
risers, 97–98, 320
risk
acceptable, 344
within asset integrity, 224–227
mitigation for, 226, 333
probability of, 225–226
of transportation, 354

Rochester Natural Gas Light Company, 27–28
rock saw, 318
root weld beads, 323–324
rotary meter, 111
routine operations, 153
rubber gaskets, 32

S

saddles, 92, 319
safety
    American Gas Association regulations for, 45–46
    of construction, 327–331
    devices for, 155
    Health and Safety Executive, 230–231, 233–234, 280
    Natural Gas Pipeline Safety Act of 1968, 179
    One Call for, 231, 236–239, 351
    pipe nomenclature for, 293
    for public, 350–354
    Safety Incidents on Natural Gas Distribution Systems: Understanding the Hazards by Cheryl Trench, 229–230
    statistics for pipelines and, 205–206
Sale of Gas Act, 25
salt dome caverns, for storage, 43, 119–120, 127, 168–169, 302
sampler, 140–141
sand padding, 318
SCADA (supervisory control and data acquisition), 150, 155
    alarm screen for, 270–271
    control systems and, 251–276
    controls compared to, 255
    front end engineering with, 284
    performance of, 271–272
    process flow for, 256–257, 273–274
scan tables, 268
scanning, 268
schedulers, 147–150, 188
scheduling
    for central control rooms, 147–150
    for gas supply management, 187–188
science
    of liquefied natural gas (LNG), 196
    of natural gas, for pipelines, 59
screwed fittings, 95, 320
sealing surfaces. *See* seats
seaports. *See* marine export facilities

seats, 101–103
secondary provers, 116, 122
sectionalizing valves, 295–296
Securities Act, 36
Securities and Exchange Act, 36
security, 273
service areas, simplified chart of, 292
service lines
    diameter ranges of, 166–167
    leaks in, 245
    plastic, 320
    plastic inside steel lines for, 242
    wrought iron pipe for, 32–33
service workers, for customer service, 181
set points, 254
shale gas, 197–198, 355
Sherman Antitrust Act, 26
ships, for transportation and storage, 200–201
shrink sleeves, as coatings, 87, 90
Simpson Natural Gas Crematory, 27
Sinclair Oil Company, 40
skelp, 84–86
skids, 79, 117
slab gate valves, 102–103
slug flow, 131–132
smart devices, 256
smart readers, 43
Smith, Adam, 335
"sniffers," 156
soap solutions
    bubbling at corrosion leak, 212
    to detect leaks, 212, 330
socket welding, 92
Soho Foundry Steam Engine Works, 16–17
solenoid actuator, 108
sound, to detect leaks, 212
source rock, 3, 14
Southern Gas Association (SGA), 244
specific gravity, 61
specific heat, 62–63
specified minimum yield strength (SMYS), 86, 121
specified minimum yield stress (SMYS), 81–82
spiral wound (steel) pipe, 85–86
springs, 106–107
squeezing off, 81, 246
stable defects, of asset integrity, 223
stakeholders, 349–350
    field operations relations with, 142
    for pipelines, 281–282

standard dimension ratio (SDR), 81, 121, 294, 317
Standard Oil Trust, 26
standard temperature and pressure (STP), 61
standards, 279. *See also* engineering standards and codes
State of the National Pipeline Infrastructure, 229-230
state-owned local distribution companies (LDCs), 160
static pressure, 68
station graphics page, for monitoring pipelines, 152
Statutory Instrument 1996 No. 825, 280
steel fittings, 91-92, 94-97, 330
steel pipe, 33. *See also* line pipe; steel fittings
    bell and spigot joint for, 32
    for construction, 320-324
    corrosion of, 117-118, 235, 242
    design of, 286-287
    hydrostatic testing of, 329
    iron content of, 235
    mains for, 164
    manufacturing improvements for, 40
    manufacturing methods of, 84-86
    for natural gas, 23-24, 27
    operating pressures for, 294
    plastic displacement of, 80-81
    plastic inside of, 242
    plastic pipe compared to, 295
    probability of failure for, 226
    properties of, 63, 86, 91
    skids made from, 79
    for storage, 119-120
    for transmission lines, 38
    welding of, 323-324
steel slab gate valves, 102
stems, 103, 105-106
Stokes, George, 62
storage
    aquifers as, 119-120, 127, 168-169, 302
    depleted reservoirs, 119-120, 127
    floating regasification units for, 202
    of gas, 40-43, 119-121, 127, 168-169, 198, 302
    line pack for, 121
    of liquefied natural gas, 199-201
    of natural gas, 13, 76, 120, 127, 302-303, 355
    purchasing capacity and rates for, 186
    reservoirs for, 3, 168-169, 302
    salt dome caverns for, 43, 119-120, 127, 168-169, 302
    of supplemental gas, 168-170
    wells for, 134
straightening vanes, for inferential meters, 111
strain, 63
straining, 81-82
stranded gas, 136-137
stressing, 81-82
stringer weld beads, for welding, 323-324
stub ends. *See* fittings
subject matter experts (SMEs), 175
supervisory control and data acquisition. *See* SCADA
supplemental gas, 168-170
supply mains, 162-163
supply redundancy chart, 171
survivability, 226
swing check valves, 103-105
system velocity, 291-293

T

tags, 265
tank train, 26
tank wagon, 26
tankers, 8, 201
tanks, 199-201
tapes, 87, 89-90
taps, 92, 166, 168, 319-320
taxes, 339
technical services, 175-176
technicians, 174-175
    for leak detection, 215, 217, 219
    maps for, 238
tees. *See* fittings
temperature
    measures and dimensions of, 60-61, 109, 139-140
    types of, 60, 66, 196, 290
Tennessee Gas Pipeline, 46
Texas Eastern Transmission, 37
therm, 61
thermodynamics, 67
thermoformed plastic fittings, 92-93
Thomson, William. *See* Kelvin
throttling, 98
time stamping, 267-268
time-dependent defects, 223
time-independent defects, 223
tolerance zone, 238-240

the tong gang, 39
tongs, 38–39
town gas. *See* manufactured gas
town gates, 13
town stations. *See* city gates
town taps. *See* city gates
tracer wire, 119
training, 156, 180
trains, for liquefied natural gas, 197
transition risers, 97–98
transitional turbulent flow, 69, 71
transmission lines, 2
    companies for, 146
    construction from 1900 to 1950, 38–40
    line packing for, 170
    linear, 287
    for processing plants, 135–136
transportation
    capacity for, 186, 355
    of crude oil, 5–8, 26–27, 37
    Department of Transportation, US, 206, 229, 237
    of liquefied natural gas, 200–201
    Oil Transportation Association, 138
    risks of, 354
    schematic, for purchasing, 185
    Transportation of Natural and Other Gas by Pipeline: Minimum Federal Safety Standards, 280
transportation rates, 199–201, 355
    billing procedures for, 336–337
    negotiations of, 190
    requests for proposals, 186
Trench, Cheryl, 229–230
trenchless construction, 43, 320–322
trucks, 5–6
turbine meters, 111–112, 139
turbo expansion. *See* refrigeration technology
turbulent flow, 70–72
turns, 317–318
Twain, Mark, 1
two-way feeds, 170–172, 242–243
Tyson Engine Company, 23

U

ultimate capacity of flow rate, 74
ultrasonic meter, 114, 139, 145
ultrasonic testing, 330
unaccounted for gas, to detect leaks, 212
unbundled rates, 336–337

underground leak detection, 219–220
underground storage, 198
uninterruptible power supply (UPS), 308–309
United Kingdom
    Gas Council of, 38
    Health and Safety Executive of, 230–231, 233–234, 280
    Pipelines Safety Regulations of 1996 for, 280
    Statutory Instrument 1996 No. 825 for, 280
United Natural Gas Company, 28
unpacking the line, 76, 148–150, 152
update frequency, 268–269
update process, 269
upper explosive limit (UEL), 210
upstream, exploration and production of, 4
US Fuel Administration (USFA), 36
utilities
    Chicago Utility Alert Network, 351
    commissions for, 36
    Joint Utility Locating Information for Excavators, 351

V

vacuum excavation, 322
valuation, 344–346
value chain, 4–5, 202
    natural gas, 14, 128
    refined products pipeline, 8–9
valves
    aboveground district pressure regulator with backup monitor valve, 165
    cast iron, 28
    common types of, 99–106, 162, 240–241, 254, 295–296
    control, 254, 266
    excess flow, 167, 241
    functions of, 98–105
    Jules-Thompson, 194
    origin points for, 265
    set points for, 254
    stems for, 103, 105–106
    technician repairing a pressure control valve, 174
Van Syckle, Samuel, 27
Vanderpool Pipeline Engineers Inc. Design Standards, 306–307

vapor. *See also* water, vapor method
    clouds, from leaks, 210
    pressure, inside natural gas, 63, 66–67
vegetation, to detect leaks, 211
velocity, 288, 291–293
venting, 210, 213
viscosity, 61–62, 111, 291
volume, 109, 290–291
volume meters, 109–111
volume provers, 116
volumes, for revenues, 357
volumetric charges, 182

W

Washington, George, 15
water
    municipal systems use of pipelines, 53–54
    pressure measurement in one cubic foot of, 72–73
    vapor method, 130–131
Watt, James, 16
way leaves. *See* easement
weather, 187, 357
weighted average, 338
welding
    butt, 93, 316–317, 320
    for facilities, 325–326
    as failure mechanisms and forces, 234–235
    girth, 87–89, 91, 330
    for maintenance precaution, 155
    for pipes, 84–86
    raised face weld neck flange, 95–96
    for repairing defects and failures, 246–247
    socket, 92
    steel, 97, 323–324
    for transmission lines, 38–39
wells, 134
Westinghouse, George, 24
Winsor, Fredrick Albert, 17
winter demand, 186
withdrawal rate, 120
wood for energy needs, 30
wrought iron pipe
    corrosion of, 235
    removal of, 286–287
    for service lines and connections, 32–33
    for transmission lines, 38

X

x-ray testing, 330

Y

Yellowstone Pipeline, 9

Z

Z factor, 64, 66
Zeebrugge Hub, 301